SCIENCE AND ENGINEERING POLICY SERIES

D1744275

**General Editors**  Professor Sir Harrie Massey
Professor Sir Frederick Dainton
The late Lord Jackson of Burnley

J. E. Mortimer

# Trade Unions and Technological Change

WILTS COUNTY
AR
LIBRARY

1 342728 000

OXFORD UNIVERSITY PRESS 1971

Oxford University Press   *Ely House, London W.1*

| | |
|---|---|
| Glasgow | Bombay |
| New York | Calcutta |
| Toronto | Madras |
| Melbourne | Karachi |
| Wellington | Lahore |
| Cape Town | Dacca |
| Salisbury | Kuala Lumpur |
| Ibadan | Singapore |
| Nairobi | Hong Kong |
| Dar es Salaam | Tokyo |
| Lusaka | |
| Addis Ababa | |

© OXFORD UNIVERSITY PRESS 1971

25. NOV. 1971

Printed in Great Britain by
William Clowes & Sons Ltd.
London, Colchester and Beccles

# Preface

This book seeks to examine a number of related questions. Firstly, what is the attitude of the trade union movement towards technological change? Secondly, how far and in what ways does the movement seek to influence change? Thirdly, what are the implications of change for the movement itself, including its structure and its functions?

Technological change cannot be considered in isolation from the structure of industry. Part of this book, therefore, discusses trade union policy on industrial structure and the measures needed to bring about changes likely to lead to the more effective use of productive resources.

There may be some who approach the subject of trade unions and technological change with their eyes mainly on restrictive practices. I hope this book will help to show that the real content of the subject is very much wider. Nevertheless, labour practices do form part of the subject and they are discussed in this book in relation to the circumstances in which they arise.

It is fair that the reader should know the basic standpoint from which the book was written. It was written in the conviction that technological change leading to greater productivity can, with accompanying measures for full employment and social welfare, provide the basis for higher living standards. Change along these lines is, therefore, desirable.

This book was written in the equal conviction that through trade unionism workers can influence the conditions under which they are employed. Trade unionism thus expresses not only an economic interest but, even more important, the desire of working people to share in the regulation of their working environment. This does not mean that trade unions commit no wrong or that all actions taken in the name of trade unionism are justifiable. But it does mean that they are a vital expression of democracy. The case for democracy is not that people and representative organizations are infallible. It is that citizens should have the right to speak in their own interest and to influence the way in which society is conducted. If this is denied, grievances will remain unresolved and may lead to revolt; mistakes will be compounded and may lead to calamity. Democracy provides a mechanism through which changes can be made grievances redressed, and mistakes rectified without unnecessary suffering.

It remains only to add that the views expressed in this book have been formed largely as the result of practical experience illuminated by the literature of trade unionism. The practical experience includes some early years of workshop and drawing-office employment in the engineering industry, some twenty-two years full-time employment in the trade union movement, and finally some two and-a-half years as a member of the National Board for Prices and Incomes. Needless to say the views expressed in this book are my own.

*London*
*December* 1970                                      **J. E. M.**

# Contents

# The trade union attitude to technological change

The attitude of trade unions towards technological change is based primarily on two considerations. The first is the effect of technological change on employment and the second is the effect on living standards.

Trade unions have always been concerned with the maintenance of full employment. It is not to them an academic economic question. It affects the welfare of their members. Among working class people in particular the memory of unemployment and the fear of unemployment is handed down from one generation to another. It has an important bearing on workers' attitudes towards industrial change. A worker is not likely to be impressed by the advantages of industrial development if the immediate consequence to him is redundancy in his existing job followed by prolonged unemployment.

But it is not only manual workers who are influenced by the fear of unemployment. It affects also other employees. Clerical workers may be apprehensive about the consequences on employment of the introduction of computers; draughtsmen may express fears about new techniques for the preparation and reproduction of drawings; scientists or technicians may feel that their employment has been so highly specialized that their skill will be of little value to other employers; and managers may dread a merger, take-over, or rationalization of company structure because they may find themselves out of a job at an age when it is difficult to find other employment with comparable status and salary.

These fears are not based only on the memories of the years of mass unemployment. They also reflect current experience. The pace of industrial change is quickening. Some industries, notably coal and cotton textiles, have declined fairly rapidly and are still contracting. Hundreds of thousands of workers have been directly affected. In the case of the mining industry many of the workers have had to leave their traditional homes. Industrial change for them has meant not only finding a new job, but uprooting their families, moving to a new area, and establishing virtually a new way of life.

Millions of workers have also been affected by the company mergers and take-overs of recent years. Changes in industrial structure in the 1960s were more rapid than ever before. More recently there has been a slowing down in the merger movement but even now there is not a month goes by without news of new take-overs and mergers. Some of these have led to the closing of factories, but even where they have not they have often been followed by reorganization. These changes have an unsettling effect on the employees. Mergers and take-overs generate a feeling of uncertainty even among workpeople whose continued employment seems assured.

Nor should it be overlooked that even in recent times there has been—and still is—an unemployment problem. In the winter of 1970 there were more than 750 000 workers out of a job. Some of this number were workers temporarily out of work who, more strictly, could be defined as moving from one job to another. Nevertheless, in certain areas of Britain unemployment was and remains a serious problem. In the development areas the unemployment rate was well above 4 per cent. In some urban districts the unemployment rate was even higher: in Sunderland and Bishop Auckland, for example, it was around 6 per cent, and in Northern Ireland it was more than 7 per cent.

### The trade union case for full employment

The attitude of the unions towards technological change rests, therefore, first and foremost on a demand for full employment. It is always unrealistic to discuss union reactions to industrial change without regard to the effect of change on the employment prospect of workers. This does not mean that the trade union movement takes the view that workers should never be expected to change their jobs. Clearly this would be impracticable and would imply that the pattern of employment and the distribution of workers between one industry and another should never alter. There will always be some industries expanding and others contracting. In a dynamic industrial society these changes will be extensive, taken over a period of years. In contrast to coal mining, cotton textiles, and agriculture, where employment has declined steeply, there are other industries and services, including engineering, building, printing and paper, chemicals, road and air transport, distribution and commercial services, where employment has expanded.

Change must take place within a context of full employment. This is the starting point for the unions. The responsibility for attaining and maintaining full employment is one that falls squarely on the government of the day. In post-war Britain the task of maintaining full employment has been closely related to two other problems, namely the balance of payments and regional development.

The periodic deficit in the balance of payments has led successive governments to take deflationary measures, sometimes described as 'stop–go' policies because of their effect on the economy. The purpose of these measures, as governments saw them, was to apply a brake on economic activity on the domestic market so that imports would be reduced and firms would bend their energies towards an expansion of exports.

'Stop–go' policies have been opposed by the trade union movement. The unions called for alternative policies to meet the deficit in the balance of payments, including a reduction in outflow of capital for investment purposes, the drastic cutting of overseas military expenditure, and the introduction of import quotas. The trade union movement argued that if these measures had been taken it would have been possible to secure a surplus in the balance of payments, maintain full employment and devote additional resources to expansion.

The unions have always insisted that, just as a policy for full employment is essential for co-operation in technological change and for overcoming traditional fears of insecurity, so, equally, special measures for regional prosperity are essential for full employment. The unions have never accepted that migration from areas of unemployment to areas of prosperity should provide the sole or even the main answer to regional difficulties. There are, of course, social costs to be borne when special measures are taken to attract new industries to the development areas. But there are also heavy social costs to be borne when a region withers because of the decline of its traditional industries. The further concentration of industries in the

Midlands and the South-East also entails social costs. Migration provides only part—but not the main part—of the answer to these problems.

The unions have supported many actions taken by successive governments to foster economic growth in the development areas. Nevertheless, their support has not been uncritical. Their main criticism has been that governments have been prepared to offer a range of inducements for regional development and then to rely on the profit motivation of individual companies to respond to these inducements. The result has been uneven. If new industries are to develop in a region, groups of firms with related products have to be attracted. The unions have urged that financial assistance should be more selective, and should be based on regional development plans prepared by the Regional Planning Boards.

**Industrial growth and living standards**

If full employment is the first requirement for the full co-operation of workers in technological change, the second is that they should share in the benefits of industrial progress. Their right to share should be recognized by all concerned, including government and employers.

There can be no serious doubt that in advanced industrial countries the development of productivity provides by far the firmest base for long-term improvements in workers' living standards. Over the last hundred years, for example, there has been a very substantial rise in living standards in Western Europe and this has been directly related to rising productivity. It has been estimated that in the period 1860–1960 *real* wages increased fourfold in France, Germany, and the United Kingdom, more than fivefold in the United States, and more than sevenfold in Sweden (Brown and Brown, 1969). (The very high rate of advance in Sweden was due partly to the low initial level of Swedish real wages in 1860 on the eve of industrialization.) Moreover this increase in real wages has been accompanied by shorter hours of work and longer holidays.

Recognition of the long-term relationship between productivity and living standards does not imply that there has been, and is, no scope for the redistribution of income, and particularly of wealth, to the advantage of workers. But it is rising productivity that has been by far the main source of rising living standards.

4

## The role of trade unionism

It would, however, be wrong to assume that technological change and innovation automatically bring better conditions for workers. They make better conditions possible, but the rewards of industrial progress are not usually distributed by employers as an act of benevolence. In many situations collective bargaining is a blunt instrument for bringing about a redistribution of income between capital and labour. Employers are often able to pass on higher costs, including labour costs, in the form of increased prices. Trade unionism has, nevertheless, enabled workers to maintain their share in the rewards of technical progress. This has been achieved by a combination of pressures exerted in collective bargaining, in the field of industrial and economic policy, and in politics.

It is no part of the argument for trade unionism to pretend that the development of capitalism has not made possible over the last one hundred years a comparatively rapid rise in the economic standard of life of the great majority of working people in the advanced industrial Western countries. Technological change to expand the flow of goods and services in society can be made to serve workers' interests. The real task for trade unionists is to co-operate in these changes whilst at the same time ensuring through collective bargaining and by other industrial, economic, and political means that those who are affected by change are protected and that the benefits of change are widely distributed. The cost of protecting those who are directly and primarily affected by change should be regarded as a first charge against the benefits of industrial expansion.

The role of trade unions in relation to technical change is, however, not only a protective one. Trade union pressure for better conditions can act also as a spur to management. In the past too few firms in Britain were pressed to utilize their resources more efficiently so that higher wages could be paid without increasing prices. The development of productivity bargaining, however, represented a significant change of emphasis. It showed recognition of the need, wherever possible, to relate higher wages to increased efficiency in the use of resources. More recently, following the effects of devaluation, rising prices, and some higher charges for certain social services, the emphasis in wage negotiations has again shifted away from productivity. The rise in the cost of living, arguments about comparability, and the needs of low-paid workers have become more widely quoted in support of wage claims. All these factors are important to unions, but the underlying problems to which they point can more easily be resolved if productivity is rising.

# Industrial structure and control

Rising living standards for workers must rest ultimately on the growth of productivity. This, in turn, will depend on the manner in which economic resources are organized and exploited. Changes in technology, the application of new tools and equipment, the provision of more efficient services, and the development of higher skills in organizing resources all contribute to greater productivity. They enable more to be produced or better services to be provided with the same input of labour.

Change and development are, therefore, essential for rising living standards. They provide a setting for the realization of trade union objectives. In the words of an American trade union leader, Mr. P. L. Siemiller, the President of the International Association of Machinists and Aerospace Workers, 'one of the first responsibilities of collective bargaining is to make technological changes socially and economically digestible'.

## Change is necessary

There are many sections of British industry where change is necessary. Some of the main weaknesses can be readily distinguished. First, the buildings, equipment, and tools are not as modern as they should be. Examples readily come to mind; engineering factories with old machine tools; clothing factories in dilapidated buildings and with sewing machines that were manufactured many years ago and do not include modern devices; building firms with an almost total lack of modern materials handling equipment; wholesale markets —for example most of Smithfield—housed in structures erected in the last century and with very little mechanical power to help convey produce from one point to another; and textile mills equipped with old machinery. The need is for a higher level of capital investment.

Secondly, there is the industrial imbalance of much of Britain's research and development. In some sectors there is too little research. In others the research and development is heavily concentrated on projects associated with either defence or aerospace. In still others, the right ratio between research and development has not been achieved. The result of this imbalance and its inadequacies is that Britain has fallen behind in some industries in product development. There are significant exceptions, for example in the development of synthetic fibres, but these exceptions are too few.

Third, the quality of marketing is not always as high as it should be and with some firms it is of a very low standard. It has been a commonplace criticism of Britain throughout the post-war period that too few firms have pushed their wares energetically in export markets and that some exporters have shown an indifference to customers' requirements or have provided poor servicing facilities. Again there are significant exceptions and the exceptions are not all to be found among the large firms whose names are familiar to everyone. Nevertheless, Britain's export effort is highly concentrated. Sir Val Duncan, Chairman of Rio Tinto Zinc, in a lecture in 1969 sponsored by the Institute of Directors, pointed out that 50 per cent of Britain's entire export business was conducted by 120 companies. The total number of companies in Britain, he said, was more than half a million.

Fourth, the quality of management in many British firms needs to be improved. There has been a strong tradition in Britain against the entry into industry of the brightest young men and women from universities, and Britain was comparatively late entering the field of management training. Big strides forward have been made in recent years, and it is not a criticism of the British Institute of Management or of the new business and management training schools to say that much still remains to be done. Even with existing resources and with little injection of new capital significant improvements in efficiency could be secured in many companies by better management.

Fifth, the labour of British workers is used much less effectively than it could be. This is partly a criticism of the educational system—with the traditionally subordinate status of technical studies—but it is even more a criticism of the traditional training methods, or lack of training, by many firms. The talents of women and girls in employment are often grossly under-utilized and their potential contribution is never realized because opportunities are not provided. Practices which are inefficient in the utilization of labour have developed in some firms or sectors of industry. Where these practices have been fostered by workers themselves it has invariably been as a protection against insecurity. This is part of the price that Britain has paid and is still paying for unemployment.

**State intervention**

The need for greater efficiency, policies of full employment and regional development, and changes in industrial structure brings to the fore two main questions to which the trade union movement has had constantly to address itself. The first is to what extent can these changes be brought about by the free operation of market forces; or, to put it another way, to what extent is the intervention of the state necessary for the achievement of maximum economic welfare? The second is what are the benefits and possible dangers that can be expected to flow from changes in industrial structure and, in particular, from the concentration of resources in larger units and the concentration of power in fewer hands?

To both of these questions the answer of the British trade union movement has been provided in a succession of policy declarations extending over many years. There have been variations in emphasis from time to time, but the main trend has been remarkably consistent. The trade union movement has been a consistent advocate of state intervention in economic affairs. In its report on post-war reconstruction prepared in 1944 the T.U.C. said that full employment, price stability, the protection of the people either as workers or consumers against exploitation, the equitable distribution of income and economic opportunity, and the promotion of national development and security were all a responsibility of government and demanded more measures of planning, regulation, and control if they were to be fulfilled. The modern economic system, argued the T.U.C. report, bore little resemblance to the capitalism of a century ago. Modern capitalism was subject to a considerable amount of control, exercised in many cases by private individuals and sometimes in a manner in which public responsibility was neither clearly defined nor accepted. It was now clear, said the T.U.C. that the liberty of the individual was most endangered by a system of unrestrained private enterprise.

The T.U.C. described as fallacious the claim that only free private enterprise could provide a rising level of industrial efficiency. On the contrary, technology had been hampered on the one hand by the lack of industrial co-ordination and on the other by the restrictive policies of private monopolies. Change, said the T.U.C., was essential for industrial efficiency. One of the strongest arguments for the transfer of key industries to public ownership and for the introduction of other forms of public control was that they would provide a framework for efficient industrial organization and for ensuring that industrial efficiency served its proper purpose of improving the standard of life of the community. The T.U.C. put the essential issue as they saw it in the following sharp terms:

The choice before us is not between control or no control, but, in principle, between control by public authority responsible to the community, or control by private groups and persons owing a final responsibility to themselves alone and, in detail, between degrees of control and types of control.

This basic conception of the need for state intervention in the conduct of industry in order to promote development and efficiency and to protect the public interest found expression in a succession of T.U.C. policy statements throughout the post-war period. They included reports on the public control of industry in 1950 and on public ownership in 1953, statements in the second half of the 1950s on the need to stimulate the economy, evidence on trade union policy for economic growth submitted to the Royal Commission on Trade Unions and Employers' Associations and, finally, the T.U.C. economic reviews—each published as a separate booklet—for 1968, 1969, 1970 and 1971.

### The concentration of ownership

One particular feature of industrial change of special interest to the trade union movement is the concentration of ownership and control of industry. The process of concentration has accelerated in recent years with more and more mergers and take-overs (Monopolies Commission, 1969). It represents a structural change which technological change has helped to bring about.

There are many reasons for mergers and for the continued concentration of the ownership and control of resources. One compelling reason, particularly in manufacturing industry, is provided by changes in technology. New and more economical methods of manufacture may depend on the introduction of costly machines. The profitable use of such machinery may, in turn, depend upon shift working and a large continuous output. The merger of two competing smaller firms may make it possible to introduce new methods and new machinery and to exploit them profitably.

Increased size may also bring advantages in other directions. It may make it possible to devote more resources to research and development, to strengthen servicing facilities, and to improve marketing arrangements. The merger of two firms with related, even if not overlapping, interests may enable the top management to rationalize the effort and resources of the two organizations. One division may concentrate on one range of products and the other division on another. A large firm may also find it easier to raise capital if it should be in need of finance to extend its operations.

Another reason for mergers is the varying qualities of management. A firm with an efficient and thrusting management may seek to extend its area of operations. The acquisition of another company may provide it with new opportunities. Conversely an ailing firm may be disposed to accept an offer of a merger from a company with a record of success.

Firms may seek also to extend their control backward to the sources of their supply or forward to their market outlets. Unilever, for example, was originally based on the business of Lever Brothers. Their interests in soap and margarine were extended to a wide range of other household products. Later they acquired a chain of retail shops and established plantations overseas to ensure supplies of raw materials for their products. Courtaulds is another example of a firm which has extended its interests both horizontally and vertically. It has taken over other textile firms, has become a world leading producer of artificial fibres, and has extended its interests forward into the clothing industry.

The strength and vigour of international competition has been an important factor in a number of mergers. British firms have found it necessary to merge to withstand competition from giant foreign— mainly American—firms. In electronics and electrical engineering the case for concentrating the available British resources and talent into fewer, more powerful companies was an extremely strong one. English Electric, Plessey, International Computers and Tabulators, and the Government were all involved in the formation of the new computer company, International Computers Ltd. The former separate British firms were dwarfed by the American giant, I.B.M.

In motor vehicle production the formation of British Leyland Motor Corporation Limited was of great importance for the British-owned sector of the industry. Three other principal producers in Britain, Ford, Vauxhall, and Chrysler (formerly Rootes), are under American control. Similarly in the ball-bearing industry there was a real danger that unless a number of firms could be brought together the British-owned sector might disappear. This merging of British firms took place in the spring and summer of 1969.

Though industrial concentration can often provide a framework for the better use and development of resources, a merger may sometimes be unfavourable for efficiency. Moreover, even a merger which has been prompted by a desire for greater efficiency may subsequently act in a manner which, though it protects its own interest, is harmful to the public interest.

Structural change in industry, leading to the formation of larger organizations, frequently affects market strength. And it is the ability of firms to influence the prices of their products by the scale of their output that provides a measure of monopoly power. Monopoly power, in this sense, is not confined to a single firm which dominates a particular market to the exclusion of all others. It exists wherever a firm 'has significant room for manoeuvre in its price or output policies, whereas other firms without such power are constrained by forces external to themselves' (Evely and Little, 1960).

Monopoly power may be exercised not only by a firm which is dominant in its own market, but by a group of firms which between them are able to regulate prices or output or both. Thus monopoly power may be exercised by a group of firms, each of which takes note—but without formal collusion—of the price behaviour of the others. It was Sir William Beveridge who pointed out in his book *Full Employment in a Free Society* that there are 'constant efforts of all private enterprises to escape in one way or another from the pressure of competition and to organize barriers against it'.

One of the main dangers of monopoly power is that it can dampen initiative and enterprise. The firm or firms concerned may look for a comfortable existence, with good profits and stability, by regulating the scale of output to suit their price policy and by protecting their products against competition. Firms may act in this way even though they do not consciously pursue a restrictive policy. Nevertheless, the result may be inertia and even stagnation. One way in which a firm with monopoly power may retard enterprise is by being unresponsive to technical progress. Each new development may be seen more as a threat to an existing comfortable situation than as an opportunity for expansion. This is not to say that firms deliberately suppress inventions in order to safeguard a monopoly. The real charge is that because of the circumstances in which they exist they may lose any strong incentive to search for change. Of course, they have to bear in mind that the economy is not stagnant and that a new and thrusting competitor, or a substitute product, may appear on the scene. This danger, as they see it, can never be completely avoided.

Firms exercising monopoly power may also seek to protect themselves from existing or new competitition by establishing an exclusive market for their product. Such a market may not always be based on the real competitive qualities of their product. On the contrary, it may be the result of intense advertising designed to induce preferences in consumers in favour of one product as against another. Protection from competition may also be sought by the development of special arrangements for distribution, including tied outlets, or by temporary selective price cutting and cross-subsidization from one product to another in an attempt to undermine the position of any would-be competitor. A firm with monopoly power may also be in a position to control one or more vital supplies essential for a would-be competitor.

When a new large organization is formed as a result of a merger, or even when a large organization evolves by its own enterprise, it is always possible that a central bureaucracy will develop. This bureaucracy may stifle initiative and create rigidities leading to higher costs and inefficiency. Large organizations also have problems of communication. Decisions taken by top management may not be transmitted clearly to lower management. Workpeople may be unaware of management intentions and there may be rumours and speculation damaging to morale. In large organizations there is always a tendency to refer matters for decision to 'higher authority'. The process of decision taking slows down. Every large organization finds it necessary to devolve authority on certain issues if efficiency is to be maintained.

Conglomerate mergers—mergers between companies with no related trade interest—may bring particular dangers to the public interest. The motivation for a conglomerate merger may have little or nothing to do with the more efficient use of resources. A firm may undertake a conglomerate merger in order to make a safe investment rather than to run the risk of innovation in its own field of activity. A conglomerate may also be able to affect competition in a number of different markets, or it may feel able to insulate itself more effectively against competitive pressure for price reductions in one of its traditional markets.

It does not follow that all conglomerate mergers are harmful. It is sometimes desirable that a firm should diversify its interests, particularly if its main assets are in a declining industry. A conglomerate merger may help a firm to face certain risks which it might otherwise find overwhelming.

## The international company

There is one kind of monopoly development which has taken on a new dimension in recent years. This is the growth of the international company. The activities of these companies span many frontiers. It is not only that their products are exported, but, much more important, they own manufacturing units and selling outlets in many countries. The economic and political significance of these international companies has grown rapidly in recent years. They will become even more important in the future. Most of them, though by no means all, are under American control.

The existence of the giant international companies gives rise to a number of special problems. An international company is answerable ultimately to its controlling shareholders. Their wishes and their interests will not always coincide with the policies and needs of the various countries in which their investments are located. When there is a divergence, whose decisions should prevail? If the host Government seeks to enforce its will in the interest, say, of full employment, the balance of payments or regional development, the foreign investor may take consequential decisions which may appear to him, in another country, to make good financial sense, but which run counter to the policies of the host Government.

An international corporation may, for example, decide to concentrate its output of certain products in one particular country or to subordinate its production and marketing facilities in one country to those of another. The countries concerned could be Britain, Germany, and France. Clearly such decisions might be unacceptable to the governments of at least two of the countries affected. Problems of this kind have already arisen not only among the countries of the Common

Market, notably France and Belgium, but they have been discussed in Britain in relation to motor vehicle production, computers, and the role of Philips of Holland as a manufacturer of electrical goods. The international companies are specially strong in industries associated with modern technology, including electrical engineering and electronics, motor vehicles, computers, and chemicals.

The late Walter Reuther, then President of the United Automobile Workers of America, many of whose members worked in the American base plants of giant international corporations, described the growth of these corporations as the dominant economic characteristic of the 1960s. In June 1969 he said that this development represented a very grave danger to the world labour movement. The international corporations were subject to no clear sovereignty. They were capable of transferring production facilities from one area of the globe to another and were normally responsive only to the imperative of maximising profits.

Reuther pointed out that the unions, which were often strongly organized in the main industrial countries, had nevertheless not developed a means of pooling their strength to match the power of the corporations on such matters as wages, prices, and the location of industry. He called for new international standards which would spell out the obligations of international companies to their workers in all countries, and urged unions to build 'an international source of worker power commensurate with the global power now uncontestedly in the hands of the macro-corporation . . .'

The view of successive British Governments on foreign investment in Britain has been that it ought generally to be encouraged. It can bring new employment (tens of thousands of jobs have been created since 1946 in Scotland by American firms), provide much-needed capital to revitalize firms which might otherwise go into sharp decline or even bankruptcy, gain access to advanced American technology for British located firms, and sometimes introduce higher standards of management. Governments have, however, occasionally sought assurances on certain issues from foreign firms before giving their sanction for a British firm to pass under foreign control. Under the Exchange Control Act, 1947, any proposal which involves a British company passing into the direct or indirect control of a person resident outside the United Kingdom requires Treasury approval.

In its *Economic Review* 1970 the T.U.C. said that 'The problems which foreign direction presents for trade unions is that where the power centre of a corporation cannot be properly identified, it is difficult to deal with.' There is thus a real diminution of job security when it is known—if indeed it is known—that discussions on jobs and production policy are taken much further away. This is sometimes a particular problem for executives, scientists, and technologists who are concerned about the location of further research and development and the recruitment of future top management. The T.U.C. suggested that the international organizations, such as O.E.C.D., E.E.C., and E.F.T.A. should collect information on the operation of international companies, covering employment, accounting procedures, capital flows, tax payments, and other matters. This would enable governments collectively to establish guidelines so that world trade could be conducted under clearly defined international rules for monopolies, mergers, and restrictive practices. The rules would also provide for controls over actions which tended to cause instability in trading patterns and world currency arrangements.

In October 1970 the T.U.C. convened a conference of affiliated unions to consider the impact of the growth of international companies on the public interest and on the interests of their employees. A series of 'targets' and 'initiatives' were proposed to provide a strategy for matching the power of trade unionism to the power of international companies.

**Trade union policy towards monopoly**

The economic and political problems which can arise from the existence of giant international corporations is, however, but one aspect—even though an extremely important aspect—of the power possessed by all large private industrial and commercial organizations. This power may affect national economic objectives in many ways. The willingness of large firms to undertake investment and thus to contribute to growth and employment depends in the end on their assessment of profit expectations. Company costs and company benefits for a particular contemplated project may differ from anticipated social costs and social benefits. Government economic planning, in an economy which is predominantly under private ownership, cannot be other than limited in scope. The Government is responsible for only some of the decisions which affect national economic performance. Many others are made by individual enterprises.

The Government's much more direct control of the public sector of industry provides it, however, with important economic levers, the effect of which when used can be felt throughout the economy. The annual investment programme of the nationalized industries in Britain is, for example, much the same as for the whole of manufacturing industry. All proposals for major price increases in the nationalized sector were for a period under the Labour Government referred for examination to the National Board for Prices and Incomes. In discharging this responsibility the Board, in effect, conducted periodic efficiency studies of sections of nationalized industry.

The policy of the trade union movement towards industrial integration, concentration, and monopoly was defined in the T.U.C. statement, *Trusts and Cartels*, approved by the 1945 Blackpool Congress. This policy has been reaffirmed and developed in a number of subsequent T.U.C. declarations. The 1945 statement said that the formation of giant industrial concerns resulted from modern technological conditions and from the competitive struggle itself. Even in the United States, where anti-trust legislation had existed since before the end of the last century, the growth of large combines had not been prevented. On the contrary, they had probably grown more rapidly and to a greater size than in Britain.

The T.U.C. statement argued that it would be both futile and undesirable to arrest the trend towards industrial concentration. The new large groupings often yielded economies in production and administrative costs, and generally observed better conditions of employment for their workpeople than small firms. Because of their desire to avoid industrial disputes which would interfere with the working of their complex organizations they were also more keenly interested in the institution and maintenance of proper machinery for collective bargaining.

The T.U.C. pointed out, however, that these advantages were not a reason for ignoring the need for public regulation to ensure that large combines were not able to use their dominant position in an industry to exploit the community. Some of these combines were in industries which ought to be nationalized. For others, some form of public supervision over prices was essential. The T.U.C. rightly anticipated many of the problems of effective price control. Their statement added:

Although price control seems a clear enough theoretical solution to the problem, it presents considerable practical difficulties. . . . The solution would seem to lie not in fixing maximum prices for all products sold by the large combines but by empowering a public authority to investigate costs and profits and to fix maximum prices in those cases where this was considered necessary in the public interest. It may be possible in some cases to rely largely upon . . . adequate publicity to prevent the large concern from abusing its monopoly powers. . . . The value of publicity as a means of control is all the greater in the case of the large combines since they are greatly concerned with their reputation both for business and other reasons.

The general conclusion put forward by the T.U.C. was that it 'would be undesirable generally to counteract the dangers arising from private monopoly in industrial organization by attempting to restore competitive conditions'. Monopoly was based upon the concentration of production as a result of technical progress, and restrictions upon competition arose not only for the purpose of maximizing profits, but also out of a desire to bring stability in a complex industrial system. Some form of price regulation was an essential preliminary to the efficient reorganization of certain industries.

For private industry the T.U.C. urged the establishment of machinery for the public supervision of combinations, associations, and agreements which provided for price management by private interests. In accordance with this policy the T.U.C. welcomed the setting up of the Monopolies Commission in 1948. This machinery, said the T.U.C., should supplement other measures to be taken by the Government to protect the public interest. These measures should include the taking into public ownership of certain industries, the fixing of maximum prices for certain products, and the promotion of co-operative or other forms of production in competition with established monopolies.

Throughout the 1950s and 1960s the trade union movement supported the work of the Monopolies Commission and the Restrictive Practices Court. In evidence to a departmental committee set up in the early 1960s to review policy on monopolies and restrictive practices the T.U.C. urged that there should be a comprehensive enquiry, by a representative body, into the social and economic effects of industrial concentration. When the then Conservative Government and later the Labour Government elected in 1964 proposed that the Monopolies Commission should be empowered to look into mergers the T.U.C. expressed their agreement.

## The merger movement

With the acceleration of the merger movement in the second half of the 1960s trade union concern at its effect on workers grew. It found expression in two resolutions adopted by the 1968 Congress. The first, moved by the National Union of Vehicle Builders, drew attention to the increasing concentration of capital, the number of closures of productive capacity, and the detrimental effect on the livelihood of workpeople. It called for greater public control of giant corporations and urged that before closures are decided there should be the fullest consultation between government, the planning authorities, the employers, and the unions.

The second resolution differed from the first in that it drew attention to the use of public money, through the Industrial Reorganization Corporation, in the merger movement. It asserted: 'There must be public accountability when public money is involved'. The mover of the resolution, Mr. John Dutton of the Association of Scientific, Technical, and Managerial Staffs, elaborated on the theme of public accountability where public money was used. Private industry, he said, received substantial subsidies. There were the selective employment tax (S.E.T.) rebates, the regional employment premium, development area loans and grants, and the funds of the Industrial Reorganization Corporation. The Government should not accept that the financial considerations underlying mergers and take-overs should be a matter for discussion between merchant bankers and the companies concerned, 'with the consequences suffered both by the workers and by the Chancellor of the Exchequer'.

The seconder of the resolution, Mr. Roy Grantham of the Clerical and Administrative Workers' Union, was, if anything, even sharper in his criticisms. The unions' job, he said, was to see that reorganizations sponsored by the Government and financed to a very large extent by the Government were carried out with a social commitment to the people of the country and not merely with a commitment to the balance sheets of the companies involved.

The seconder of the motion was followed by Mr. Leslie Cannon of the electricians, who sat on the I.R.C. with the support of the T.U.C. as a member drawn from the trade union movement. Mr. Cannon said that unions were right to be concerned about some of the consequences of the concentration of industry and he wholeheartedly supported demands for full and adequate consultation. He pointed out, however, that the fragmentation of industry also had grave dangers. Sometimes it led to a lack of competitiveness in world markets, a lack of research, and a lack of capital resources for essential technological development.

Mr. Cannon referred also to a paper given by Mr. Frank Cousins in 1965 in which he made the point that between the present fragmented state of British industry and the state of monopoly there was immense scope for the restructuring of British industry. The formation of the I.R.C. was one of the most imaginative pieces of legislation to come from the Ministry of Technology.

The *Economic Review* 1969 of the T.U.C. contained an extensive section dealing with mergers and their effect on manpower. The General Council accepted that it was in the interests of trade unionists to work with and not against the forces of innovation, but this did not mean that trade unionists accepted the values and standards of the market. Nor did they accept that employers and financiers were entitled to shape the development of the country without being subject to democratic pressures and without accounting to the community.

The T.U.C. pointed out that many mergers did not provide a social and economic benefit to the community equivalent to the financial benefits accruing to the companies concerned. The merger movement had been haphazard and spasmodic and was strongly influenced by immediate financial and stock exchange considerations rather than by the longer term needs of industry. The direction in which a solution had to be sought, said the T.U.C., was to bring the process of company reorganization under much more adequate planned control.

The T.U.C. described the I.R.C. as one of the most successful of the Labour Government's innovations. Nevertheless, the T.U.C. did not feel that it was satisfactory to leave to the I.R.C. the job of defining the general considerations which, when assessing a proposed merger, could together be regarded as constituting the national interest. Such a definition of criteria could only be made in the context of a national development plan. It was above all a task which should be undertaken by the government, though obviously in formulating the criteria the government should draw upon the experience of the I.R.C.

The General Council put forward its own suggestions for a set of standards by which to appraise whether a particular change was likely to contribute to the productive utilization of the nation's resources. They included:

Will expenditure on investment and research be promoted in a more economical way?

Will manpower be used more effectively and will employment opportunities in the longer term be maximized?

Will there be economies of scale in relation to production, development, and marketing?

Will management be strengthened?

Will the capacity of the industry to compete in export and domestic markets be developed?

The T.U.C. called upon the Government to use existing public agencies—such as the I.R.C., the Monopolies Commission, and the Economic Development Committees—more vigorously to assure itself of the economic benefit to the community of particular mergers and to ensure that merged companies should follow a carefully co-ordinated development plan. If necessary, urged the T.U.C., more extensive power should be taken to require companies to draw up development plans in conjunction with public agencies.

The General Council of the T.U.C. also argued in favour of permanent arrangements to increase the accountability of all large firms both to the public and to their employees, to make more adequate provision for the regular scrutiny of all firms in a monopoly supplier situation, and to provide for a more thorough examination of proposed take-overs, together with systematic follow-up reviews at intervals after the take-over. It was also insisted that account should be taken of the assistance given to industry from public funds, such as investment grants and the specific sums provided by the I.R.C. In all firms in which significant amounts of public money were invested there should be public directors to serve with other members on the board of directors. When a merger was approved by the government or not opposed the company concerned should be required to provide information at regular intervals relating to its intentions in the fields of investment, output, exports, costs, and prices.

The T.U.C. attached particular importance to the effect on workers of reorganizations and mergers. They called for an agreed code of good practice which companies would be required to accept as a condition for government assistance in reorganization or approval for proposed mergers. The code, it was suggested, would require the

company to negotiate in good faith with the union or unions concerned with a view to reaching agreement covering five main areas.
The code should incorporate five provisions.

(1) Trade union recognition and adequate facilities for effective
trade union functioning, including encouragement of workers to
join unions, the check-off where desired, and time off for workplace
representatives to deal with the problems of members.

(2) Adequate machinery for negotiation and consultation, with a
procedure for the settlement of disputes and for appeals by individual
workers against dismissal.

(3) The disclosure of information to unions, where they wished to
participate in manpower planning, concerning future plans likely
directly or indirectly to affect the level of employment and wages or
working conditions. This should include information about proposed
changes in structure, location, size of plant, production, and employment policies.

(4) Adequate procedures for dealing with such redundancies as
might occur.

(5) The negotiation of adequate compensation for workers made
redundant, who suffered diminution in their income or whose pros-
pects, seniority, or status was adversely affected.

At the 1969 Congress a resolution was carried which in effect
confirmed the policy outlined earlier in the year in the economic
review prepared by the General Council. It expressed concern at the
number of factory closures resulting from mergers, and advocated
an agreed code of good practice providing proper consultation with
the trade unions and workers involved and obliging companies to
negotiate with the appropriate unions over manpower planning and
redundancy arrangements.

It was significant that in the subsequent debate a number of
speakers, all on the left of the trade union movement, recognized that
mergers were sometimes necessary for the greater efficiency of British
industry. They were, however, deeply concerned that when mergers
took place the interests of workers should be fully considered. Reorganization, said Mr. Scanlon, who moved the motion on behalf of
the A.E.F., was still largely at the mercy of market fluctuations and
not the result of planned rationalization. The one dissenting voice
came from the draughtsmen. Their spokesman, Mr. G. W. Strattan,
said his union utterly rejected the resolution's opening paragraph
which recognized the need for mergers. He called upon the T.U.C. to

issue a clear statement to the effect that redundancy agreements were no substitute for security of employment.

## New policies

Towards the end of 1969 the Labour Government announced that it proposed to merge the Monopolies Commission and the National Board for Prices and Incomes into a new body to be known as the Commission for Industry and Manpower. The case for the merger was that the functions of the two bodies overlapped. Both had on occasions examined the same firms or sections of industry. The Monopolies Commission was concerned primarily with industrial structure and the P.I.B. with price behaviour but, in fact, the two were very closely related. One of the main ways in which market power manifested itself was through prices. Firms with substantial market power were able to regulate their prices, their output, or both, in order to maximize their profits.

The emphasis of the new body was to be more on the examination of market power and prices than on wages. The Government did not propose to take power to defer the implementation of wage or salary settlements, though it was to be still open to them to refer general wage and salary questions for examination and report.

In its *Economic Review* 1970 the T.U.C. described the Government's proposals for the C.I.M. as 'timely'. The General Council of the T.U.C. welcomed the general conception of the C.I.M. and said that the previous separation of functions for investigating and intervening in the structure and functions of large firms between (among others) the Board of Trade, the Monopolies Commission, the Industrial Reorganization Corporation, the Department of Employment and Productivity, the Prices and Incomes Board, the Ministry of Technology, and the boards that might be set up under the Industrial Expansion Act was a handicap to adequate study and a co-ordinated approach. The T.U.C. suggested some additional areas in which the work of the C.I.M. should develop, including the public accountability of firms with significant market power, the examination of overlapping directorships, the operation of 'conglomerates' and the accounting practices of firms with extensive overseas interests.

When the Conservative Party won the 1970 General Election there was no chance that the C.I.M. would come into existence. Their view was that it represented an unjustifiable interference in the affairs of industry. They saw no reason why all firms with assets exceeding

£10 m. should be 'referable' to the Commission. Nor did they accept the justification of the proposals for referring a merger to the C.I.M. up to two years after the merger had taken place. The C.I.M., said the Conservatives, would create uncertainty and apprehension in industry and this, in turn, would inhibit and discourage enterprise and initiative. Conservative spokesmen had also argued that the C.I.M. proposal was heavily one-sided. Prices, trading arrangements, and industrial structure would come under scrutiny—with attendant powers of restraint—but the practices of labour would largely escape attention. The Conservative Party urged that what was really needed was the creation of a competitive environment and less day-to-day interference in the conduct of industry.

The new Conservative Government not only dropped the C.I.M. Bill but also announced in October and November 1970 that the Industrial Reorganization Corporation and the National Board for Prices and Incomes would be abolished. In the Government's view private organizations in the City of London would be able to under-take—as they had done in the past—the arrangement of industrial mergers. The N.B.P.I. would be replaced by three review bodies, with interlocking membership, dealing respectively with top salaries in the public sector, armed forces pay, and the remuneration of doctors and dentists in the National Health Service. A new Office of Manpower Economics has also been set up to assist the review bodies and to examine particular problems on pay and productivity. The Government also indicated that they would be making proposals to strengthen and develop the work of the Monopolies Commission.

To the trade union movement the weakness of the new Government's attitude towards industrial structure is that its reliance on and endeavour to strengthen competitive forces provides an inadequate answer to the problems of British industry. The inherent trend of industry is not towards competition but away from it. The trend is reinforced by a range of technological, production, marketing, financial, and other economic considerations. It is one thing to strengthen competition by the abolition of resale price maintenance and by outlawing certain restrictive trading arrangements; it is a problem of a different dimension to deal with the growing market power of Britain's larger firms.

# The unions and productivity

The interest of the T.U.C. and of the trade union movement in productivity has been consistent throughout the post-war period. The movement has generally accepted that greater productivity has had a vital part to play in Britain's economic strategy, in the expansion of exports, and in providing the basis for higher living standards. At the same time the trade union movement has never shared the view that productivity can in some way be divorced from wider questions of economic policy. The rate at which the economy expands depends substantially not on decisions taken by unions or on a resolve by workers to work harder but on the economic policy of the government of the day. It is within the power of government to stimulate or retard economic growth.

Nor has the trade union movement shared the view that, given an economic policy of expansion, productivity is dependent primarily on the intensification of manual effort. On the contrary, many of the measures likely to lead to increased productivity have the effect not of increasing but of reducing the burden of physical toil. A bulldozer, for example, can remove earth much more effectively than a navvy with a pick and a shovel, but it is the navvy and not the driver of a bulldozer whose task is made so unattractive by sheer physical toil. The main long-term determinants of productivity are tools, equipment, methods, the available mechanical power, and the application of managerial skill in organizing the factors of production.

## Productivity campaigns

At the end of the Second World War productivity committees consisting of management and workers' representatives existed in many factories. Some of them continued into the post-war period, and it was the official policy of the Government, supported by the

T.U.C., to encourage them on a voluntary basis. At national level the T.U.C. was represented with the Government and employers on a National Production Advisory Council on Industry and there were trade union representatives on Regional Boards for Industry.

The 1947 Congress adopted a resolution welcoming the setting-up of a national economic planning board, calling for the setting of two-year production targets, and urging the fullest possible publicity to show the nation that 'their standard of living depends upon the productivity of their work'. A productivity campaign launched by the Government resulted in the holding of thousands of factory meetings, lectures, and film shows and the publication of a new journal, *Target*, which was circulated to factories and other workplaces throughout Britain. Other initiatives included the formation by the Government of a Committee on Industrial Productivity, which included trade union representatives, the setting-up of a joint body, which eventually became the Anglo-American Council on Productivity and later still the British Productivity Council, the establishment of the British Institute of Management, and the holding of trade union productivity conferences.

The Anglo-American Council on Productivity, which included representatives of employers and workers in both Britain and the United States, embarked on an ambitious programme. Committees were established for the exchange of information about production techniques, and many British teams, including supervisors, technicians, and operatives, visited American factories. Studies were also made of the maintenance of plant, the measurement of productivity, and the scope for specialization, simplification, and standardization in production. The unions participated fully in these activities.

In November, 1950, after a visit to the United States by a team of British trade union officials sponsored by the Anglo-American Council on Productivity, the General Council of the T.U.C. issued a recommendation to all unions to interest themselves actively in industrial efficiency, to employ staff qualified to advise and assist on production problems, to provide generous educational and training facilities on production subjects for union members, and to publish material in union journals. The T.U.C. itself formed a production department. The response from unions and active union members varied. A number of unions sought to stimulate the interest of the rank and file in productivity and many local officials co-operated in joint activities with employers. Well over 500 000 copies of reports of productivity teams which had visited the United States were sold.

### Criticism

On the other hand, the productivity drive was running into opposition from sections of the rank and file. There had been a worsening of the economic situation, partly as a result of rearmament. The 1952 annual report of the T.U.C. acknowledged that rearmament had had an unfavourable effect on the balance of payments and that resources and manpower had been transferred to military work. There had been a consequential fall in the rate of increase of engineering exports.

To some trade unionists the question inevitably arose as to whether the productivity drive was to support rearmament or better living standards. The T.U.C., nevertheless, supported the rearmament programme. A statement prepared by the General Council said that it was an unfortunate but vital necessity. The T.U.C. statement observed that there were a few who 'because the object of their first loyalty exists outside the country' had consistently opposed 'any rearmament of the free nations'.

This did less than justice to many active trade unionists and others in the Labour movement who were critical of the scale of rearmament. There were some critics of British rearmament who were totally uncritical of the Soviet Union. But there were many others who, though opponents of communist methods and of many Soviet actions, did not regard the Soviet Union as a military threat to the West. They saw the 'cold war' primarily as a clash between the determination of the Soviet Union to maintain its domination in Eastern Europe and the determination of the United States to roll back communist control and defeat communist-influenced colonial liberation movements, particularly in Asia. A number of Labour leaders, including Aneurin Bevan and Harold Wilson, it will be recalled, were critics of the scale of British rearmament. The productivity drive, in the context of a controversial rearmament programme, did not evoke any great enthusiasm in the factories.

The controversy found expression at the 1953 Congress when the boilermakers sought to restrain the General Council from supporting the setting-up of local productivity committees. They were supported by the engineers and the sheet-metal workers. Between them these unions represented a very high proportion of the skilled workers in the engineering and allied industries, the very industries in which the war-time productivity committees had been born. In other engineering unions, including the Electrical Trades Union, the draughtsmen, the foundry workers, and the vehicle builders, there were also influential sections which were critical of the General Council's support for the productivity drive, though none of these unions spoke in the debate.

The debate was not focused on the central issue of the attitude of the unions towards productivity, and the spokesmen for the boiler-makers, Mr. Ted Hill, repudiated any suggestion that his union was opposed to productivity. Nevertheless, the debate was indicative of the lack of enthusiasm for local productivity committees to be found among some trade unionists. The boilermakers were defeated and the policy of the General Council was endorsed by Congress.

During the early 1950s the Production Committee of the T.U.C., with its energetic and able secretary Mr. Ted Fletcher, expanded the facilities for the training of union representatives in production subjects and managerial techniques. A number of firms of industrial consultants co-operated in providing training facilities and the T.U.C. established its own courses. Later a range of memoranda were published on production subjects which provided an explanation of various management techniques and offered advice and guidance to trade unionists for furthering workers' interests. These memoranda were helpful to many trade unionists.

**Automation**

In the middle 1950s a new word became familiar in the vocabulary of the trade union movement: *automation*. In 1955 the Congress adopted a resolution which said that Britain stood at the threshold of technological advance, including electronic and automatic processes, which would present the unions with new opportunities for securing higher living standards. These opportunities would, however, it said, be attended by new problems. Joint consultation and a greater measure of workers' participation were needed to solve the problems.

In 1956 the T.U.C. issued a special report on automation. Automation was defined—following a report prepared by the Department of Scientific and Industrial Research—as a merging of three main-streams of technical progress: automatic machinery, automatic process control, and the automatic processing of data. Automation, it was pointed out, was far from being a new phenomenon in industry.

The displacement of manpower was as old as industry itself. What was new about automation was the speed with which it was being extended in industry.

The T.U.C. report discussed a number of factors involved in the introduction and development of automatic processes and controls in industry and offices. It was important, the report urged, to maintain adequate capital expenditure in industry so that new equipment could be installed. More scientists, managers, and technicians would also be required. Industry would need to pay more attention than in the past to the provision of educational and training facilities. The development of automation would place a special burden on the electronic and machine tool industries. These industries would have to prove capable of producing the required new equipment. Automation would also put a premium on management efficiency. Increased attention would have to be given to production planning and control, preventive maintenance, continuous running, and the organization of supplies.

The T.U.C. report emphasized that when considering automation it was necessary to maintain a sense of perspective. Progress would be rapid in some industries but there were other industries, including agriculture, forestry and fishing, mining and quarrying, clothing, building, professional services, catering, and entertainment where the nature of the work, though offering scope for various forms of development, did not lend itself readily to automatic methods. Even in industries where automation was introduced, such as motor-car manufacturing, it would not necessarily involve all operations.

In discussing the implications of automation for trade unions the T.U.C. said there were both potential advantages and threats. The potential advantages were improvements in wages, hours, holidays, and working conditions made possible by greater productivity. Unit costs of production could be lowered offering greater stability and security of employment. On the other side were problems of labour displacement, of transfer, skill, training, wage rates, and promotion opportunities. The T.U.C. said that it was vitally important to the unions that the machinery of industrial relations should be able to deal with the problems of automation. More information on development plans and policies, including financial and cost information, should be given by management to union representatives.

The T.U.C. statement recognized that in certain circumstances redundancy might follow from the introduction of new methods and processes. It added: 'Given adequate foresight and planning, however, and taking into account normal labour turnover, the provision of suitable alternative work in the company concerned or in the area should not prove difficult within the framework of a full employment economy.'

The T.U.C. pointed out that even where redundancy did occur there were ways and means of minimizing the hardship on those affected. Compensatory payments to displaced workers were a first charge on the benefits to flow from increased productivity. Government action could also help to develop new employment opportunities in areas where traditional industries were declining. It was particularly important to provide training for displaced workers.

The 1956 Congress carried a resolution recognizing the beneficial potentialities of automation and calling for full consultation with unions where automation was being applied. The resolution also asserted that it was the responsibility of unions to ensure that the fullest consideration was given to recruitment and training policies, the avoidance of redundancy, the maintenance of earnings, price reductions, and the payment of adequate maintenance for displaced labour.

The widespread discussion in the trade union movement of the effects of automation helped to underline the need to minimize the hardship associated with redundancy. A survey of redundancy agreements made at that time showed that few agreements dealt adequately with all the main aspects of the problem, including prior consultation, selection procedure, periods of notice, compensation arrangements, and schemes for retraining and re-employment. In a large number of cases, however, there was mutual understanding between employers and unions on the procedure to be followed. There was no conclusive evidence that the provisions contained in formal agreements were generally superior to those of 'understanding'.

### Productivity, prices, and incomes

In 1957 the Government sought to give more formal recognition to the relationship between productivity and prices and incomes. The last occasion on which this had been done was in the later years of the period of office of the first post-war Labour Government. The unions, by majority vote, had then accepted a policy of restraint. This finally broke down under the impact of rising prices caused by rearmament and the Korean War. In August 1957 the Government set up an independent committee under Lord Cohen, to consider prices, productivity, and incomes. The trade union movement reacted with coolness towards it. The view of the T.U.C. was that fundamental differences on economic policy could not be removed by the setting up of such a body. The publication of the first report of the Cohen Council in February 1958 was subsequently deplored by the General Council as being of a 'partisan nature' which did little more than 'endorse the measures taken by the Government . . .'

The hostile reaction of the unions to a temporary 'pay pause' introduced by the Government in 1961 did not deter the Government from seeking support from both sides of industry for the working out of an incomes policy. The unions remained critical. The T.U.C. accepted that it was a condition of reasonable price stability that increases in incomes should keep in step with the growth of real output. They pointed out, however, that production and productivity had declined as a result of the Government's policy, and that incomes, and more particularly wages, could not be considered in isolation from other aspects of the Government's policy. The T.U.C. called for 'positive measures to secure sustained expansion'. When the National Incomes Commission was later established by the Government it was roundly condemned by the unions. It was described as 'misconceived' and 'unworkable'.

The T.U.C., however, responded much more sympathetically to a proposal by the Government that joint machinery should be established to help in national economic planning. The National Economic Development Council was set up, and the T.U.C. accepted an invitation to participate. In doing so they were, in their own words, 'well aware that the Government's proposals to set up the N.E.D.C. did not prove that the Government was committed to economic planning as the Congress understands it'. Nevertheless, they felt that trade union participation would put to a practical test the question whether participation could provide a genuine opportunity to influence Government policies in ways which would help trade unionists.

The election of a Labour Government in 1964 was followed by a renewed—but this time rather different—attempt to secure trade union support for a productivity, prices, and incomes policy. The Government emphasized that it was concerned not only with wages and salaries but with all forms of incomes, including dividends. It intended, it said, to ensure that pressure should be exerted to hold down prices as far as possible. The policy was to rest upon a programme of economic expansion. It was, therefore, according to its sponsors, a productivity, prices, and incomes policy and not merely an incomes policy.

At the outset the policy was supported by both the T.U.C. and the principal employers' organizations. There were critics within the unions but they were in a minority. They argued that a declaration of intent was no substitute for effective economic planning and that an incomes policy could only be acceptable in a context of expansion, the redistribution of wealth and income, and the drastic cutting of overseas military expenditure so that economic growth would not be jeopardized by balance-of-payments deficits.

In the event, the policy ran into serious difficulties because of the balance-of-payments problem. When it took office the Labour Government had inherited a huge balance-of-payments deficit. The solving of this problem, but with only limited changes in overseas military spending, became an overriding constraint on almost every aspect of Government policy. The economy was restrained, curbs and squeezes were imposed, and taxes were increased. Eventually, after devaluing the currency, an improvement was secured. For the first time for many years the British balance of payments ran into surplus simultaneously with expansion at home.

The strains of this period were too great for the prices and incomes policy to operate as originally envisaged. Devaluation caused inevitable price increases, and it was the Government's policy to attempt to enforce a reduction in consumer spending so that resources could be switched from the home market to exports. Union members insisted on wage and salary increases to prevent, as they saw it, a reduction in living standards. In order then to reduce the inflationary pressure of home demand the Government increased taxation thus, in effect, giving a further twist to the upward spiral of prices. All this took place during a period when, because of the fear of excessive imports, the Government were deliberately restraining the growth of the economy. This tended to increase unit costs in many sectors of industry. It was also a period of internationally high interest rates. The prices and incomes policy, as originally envisaged, could not hope to survive in such a combination of unfavourable circumstances.

Thus the prices and incomes policy, launched in 1964 by the newly elected Labour Government, came to an end. It came to an end because of the circumstances under which it had to exist. But the central problem to which it was addressed, namely how to reconcile full employment, economic growth, and price stability, remained. The attainment of the separate objectives has to take place within a framework of widespread oligopoly in industry, a strong and free trade union movement, and social demands for rising living standards, extended social services, and greater equality in society.

The prices and incomes policy did, however, for a period achieve a significant change in mood in British collective bargaining. To a far greater extent than hitherto negotiators on both sides became conscious of the need to relate changes in incomes to greater productivity and efficiency. Similar considerations were also given more emphasis in price determination. This change in mood was helped by the many reports published by the Prices and Incomes Board. These reports focused attention on the possibilities of increasing efficiency in almost all the industries and services referred to the Board for examination. This was the main constructive role of the P.I.B. It never ceased to emphasize that its concern was not just with prices and incomes, but with *productivity*, prices, and incomes.

## Productivity conferences

In the early autumn of 1966 and in the summer of 1967 the Government convened national conferences on productivity. Employers and T.U.C. representatives participated, together with ministers and specialists for the various subjects discussed. The main subjects at the first conference were the relationship between productivity and the national rate of investment; productivity techniques; and manpower utilization. At the second conference the main subjects were planning; productivity in relation to marketing and distribution, including transport; and the role of technology in productivity. Various follow-up measures were taken to assist the performance of industry. The T.U.C. and the C.B.I. also co-operated in holding several regional conferences on productivity. The scope and extent of the activities of the British Productivity Council were also increased.

In June 1967 the C.B.I. and the T.U.C. issued a joint statement on productivity. The statement made the following points.

(1) That both sides were committed to increasing output and real incomes combined with full employment.

(2) That a main aim of the joint initiative would be to promote the more efficient use of manpower. Employers should discuss with unions the more efficient use of labour.

(3) Employers' organizations and unions should report back to the C.B.I. and the T.U.C. on the progress made and any problems which might arise. These could then be discussed jointly.

(4) That in joint discussions on productivity, workers' representatives should be free to raise any matter appearing to them to have a bearing on an industry's or firm's productivity; and should expect to receive such information as was necessary to make the discussion purposive and, as would be reasonable in the circumstance of the case, to expect management to supply.

(5) That, as far as possible, negotiations on productivity issues should be carried out through established negotiating machinery.

## Technological innovation

In June 1968 the General Council of the T.U.C. sent to the Ministry of Technology its observations on a report on technological innovation prepared by the Central Advisory Council for Science and Technology. The report expressed the view that the widening technological gap between Britain and the United States was not principally due to any greater skill or lower costs in the United States but to Britain's comparative failure to establish priorities for research and development and to her wrong development of resources.

The T.U.C. statement pointed out that Britain spends more of her national resources on research and development than any other country, excluding the United States and the Soviet Union, but that a large proportion of this sum is taken up by research on military purposes. The General Council urged the Government to establish clear priorities and to give greater support for civil research.

The T.U.C. said that it was technological initiative that was lacking in British industry. This initiative represented the will and ability of industry to translate research and development into production on a commercial basis. Too few qualified men were employed in the productive processes in industry in contrast to the number employed in research.

In one of their regular periodic reviews of the effects of automation and technological change the General Council reported in September 1969 that there was an increasing use of advanced management techniques in industry, including the application of computers to a widening range of tasks. These new techniques were expanding not only in manufacturing industry, but also in various non-manual fields. The new methods were having an impact on the pattern of working hours, including shift work. Technological changes were also an important factor in the growing number of mergers in industry. Finally the General Council drew attention to the relationship between technological change and trade union attitudes, including the growing emphasis given to productivity in wage and salary bargaining and the demand for workers' participation in the conduct of industry.

## Productivity bargaining

It was at the beginning of the 1960s that the term 'productivity bargaining' came into fashion. It was intended to describe a form of collective bargaining with features distinguishable from other forms of bargaining. The main features are as follows.

(1) It relates wage or salary increases to planned increases in productivity.
(2) It seeks to promote the more effective use of all the resources of a plant, industry or service. Thus productivity bargaining requires careful planning and managerial skill.
(3) It specifies such changes in working practices, including those of labour, as are necessary for the more effective use of resources.
(4) It provides for wage or salary increases in anticipation of planned changes leading to increased productivity, or for payment of wage and salary increases as the changes take place and the cost savings are achieved.

There have been many agreements in industry described as 'productivity agreements' which do not contain all these characteristics. It is not a very helpful exercise to argue to what extent they are genuine productivity agreements. What is really significant is that the spread of productivity bargaining has encouraged employers and unions to concern themselves in their bargaining with productivity and the efficient use of resources, including the more effective use of labour.

The research paper on productivity bargaining prepared for the Royal Commission on Trade Unions and Employers' Associations described Allan Flander's book *The Fawley Productivity Agreements* as the indispensable starting-point for any consideration of the subject of productivity bargaining. This book provided a detailed account of an elaborate set of agreements concluded in 1960 between the management of the Esso refinery at Fawley and a number of trade unions. These agreements, according to Allan Flanders, 'were without precedent or even proximate parallel in the history of collective bargaining in Great Britain'.

The distinctive features of the Fawley agreements were, first, that the company agreed to pay large increases in wages, approximately 40 per cent, in return for defined changes in working practices that were hampering the more efficient utilization of labour, and, second, that provision was made for a rigorous and sustained effort to reduce and even to eliminate systematic overtime. The changes in working practices included some relaxation of job demarcations, the withdrawal of craftsmen's mates and their transfer to other work, and more shift-working. The wage increases granted under the agreements were paid in instalments to compensate for the fall in overtime pay as the programme for the reduction in overtime took effect.

### P.I.B. reports on productivity bargaining

In its June 1967 report on productivity agreements (Report no. 36) the Prices and Incomes Board defined a productivity agreement as 'one in which workers agree to make a change, or a number of changes, in working practices that will in itself—leaving out any compensating pay increase—lead to more economical working; and in return the employer agrees to a higher level of pay or other benefits'. The report pointed out that the productivity criterion first laid down in the 1965 White Paper on *Prices and Incomes* (Command 2639) and repeated in the White Paper on *Prices and Incomes Policy after 20th June* 1967 (Command 3235) permitted increases in pay above the norm:

where the employees concerned, for example by accepting more exacting work or a major change in working practices, make a direct contribution towards increasing productivity in the particular firm or industry.

P.I.B. Report no. 36 then added that this requirement excluded a mere promise of greater effort or efficiency in return for higher pay. The contribution, said Report no. 36, must be 'direct'. Secondly, a payment consequent upon greater efficiency where this was caused by technological advance alone was also excluded. The words 'for example' in the productivity criterion in the incomes policy White Papers (given in the preceding paragraph) were dropped in the P.I.B. report, which said that in a productivity agreement there *must* be a contribution from the workers by way of (not, for example, as in the White Paper) 'more exacting work or a major change in working practices'. This assertion was repeated in the P.I.B. guidelines for productivity agreements.

P.I.B. Report no. 36 emphasized that an agreement for a new payment-by-results scheme would not necessarily qualify as a productivity agreement. The essential ingredient for a productivity agreement was a change in method of operation. A new payment-by-results scheme might provide pay increases in return for higher output or better performance but need not involve a change in method of operation. The agreements referred to the P.I.B., which the Board discussed in Report no. 36, aimed at achieving various changes in working practices, including flexibility of working between different groups of workers, the reduction of overtime, removal of restrictions on output, manpower reductions, and changes in patterns of work. In return the agreements provided for higher pay and changes in pay structure.

Although the agreements provided for changes in working practices this was not the usual point from which the productivity proposals were initiated. The starting point was 'the preparation by management of plans for new and more effective methods of operation for the whole plant or company'. Only when these plans had been drawn up, said the P.I.B. report, could a company identify the needed changes in practices and estimate the possible alterations in pay. The careful preparation of estimates of the likely savings and costs of the proposed changes was an essential part of the planning programme. Once an agreement had been concluded effective controls were necessary to ensure that the stated objectives were in fact achieved.

The P.I.B. first published a set of guidelines in December 1966 to assist management and unions when drawing up productivity agreements. These were amended by Report no. 36 published in June 1967 and the amended guidelines were widely publicized. They were adopted in the Government's White Paper on *Productivity, Prices and Incomes Policy in* 1968 *and* 1969 published in April 1968. The guidelines were as follows:

(1) It should be shown that the workers are making a direct contribution towards increasing productivity by accepting more exacting work or a major change in working practices.
(2) Forecasts of increased productivity should be derived by the application of proper work-standards.
(3) An accurate calculation of the gains and the costs should

normally show that the total cost per unit of output, taking into account the effect on capital, will be reduced.

(4) The scheme should contain effective controls to ensure that the projected increase in productivity is achieved, and that payment is made only as productivity increases or as changes in working practice take place.

(5) The undertaking should be ready to show clear benefits to the consumer through a contribution to stable prices.

(6) An agreement covering part of an undertaking should bear the cost of consequential increases elsewhere in the same undertaking, if any have to be granted.

(7) In all cases negotiators should beware of setting extravagant levels of pay which would provoke resentment outside.

**Another report by the P.I.B.**

Nearly eighteen months later, in November 1968, the P.I.B. were again asked to report on productivity agreements. The terms of reference were drawn up to focus attention on some of the problems which experience of the extension of productivity bargaining had shown to need further examination.

The P.I.B. reported in the summer of 1969. The report stated that by June 1969 the register of productivity cases kept by the Department of Employment and Productivity had recorded some 3000 arrangements covering approximately 6 million workers, or about 25 per cent of all employed persons. Not all of these cases came within the definition of a productivity agreement set out in Report no. 36. Some of them were concerned mainly with the revision of wages structures and improvements to existing payment-by-results schemes. Nevertheless, many of them did provide for changes in working practices, such as increased flexibility in the use of manpower or the elimination of mates and their transfer and training for more skilled work.

The report, on the basis of its examination, drew a number of main conclusions. One of the most important was that in a substantial number of cases the net effect of a productivity agreement was the achievement of lower costs per unit of output. A second conclusion was that, in the main, productivity agreements had contributed to better industrial relations. They had led to changes which would otherwise have been likely to give rise to contention. In other words changes in working practices had been introduced in exchange for wage increases.

A third conclusion drawn by the P.I.B. was that the negotiation of productivity agreements had given a greater breadth to the concept of co-operation. Management had usually secured 'a better controlled payments system' and the more effective use of resources at its disposal. This, however, had not been obtained by the assertion of so-called managerial prerogatives but 'through the consent and co-operation of workers and through greater participation by trade unions in decision taking either through shop stewards or full-time officials'. Productivity agreements, said the P.I.B. report, had placed a premium on positive co-operation between management and workers in raising efficiency. Greater authority and responsibility had been thrust on trade union representatives at plant level and there was greater 'democratization' on the shop floor.

The report spoke emphatically of the possibility of non-manual workers contributing to greater efficiency. Despite a co-operative attitude by the workers concerned, the report asserted, managements had tended to neglect opportunities to raise the efficiency with which they employed non-manual workers. This was a major requirement to be made good.

## Why unions participate

In general, unions have participated in productivity bargaining not because it offers a solution to all problems but because it has provided a framework within which negotiators, whether full-time officials or shop stewards, backed by strong trade union organization, can secure significant improvements in wages and conditions and can extend the area over which trade union influence is exerted. This does not mean that all productivity agreements are beyond criticism. Productivity bargaining, like any other kind of collective bargaining, may lead to agreements which are deficient. There is, however, nothing new about such deficiencies in industrial relations. In the approach to pro-ductivity bargaining it is important that unions are able to develop the new opportunities which present themselves based upon strong and active trade union organization, the competent representation of workers' interests and the widening scope of bargaining in circum-stances of greater industrial efficiency.

A useful starting point for any trade union discussion about productivity bargaining is the proposition—the validity of which

had been underlined again and again in trade union experience—
that unions are more likely to secure good wages and conditions in a
highly productive, efficient, and prosperous industry or undertaking
than in one which is backward, inefficient, and depressed. This point
was emphasised in the discussion document on low pay published by
the T.U.C. in 1970. It pointed out that in certain sectors there was a
relatively narrow spread between the earnings of the lower paid and
the medium paid workers. This, it said, 'indicates that the problem is
one of low prosperity in the industry generally, that could presumably
only be solved by measures to increase overall productivity and
efficiency'.

Workers *do* have an interest in the efficient organization of industry.
The growth of national productivity depends, of course, on much
else besides the effects of collective bargaining. Unions are well
aware of the importance of these other issues and, as described in an
earlier chapter, the T.U.C. have discussed them in the annual economic reviews which they have produced and submitted to affiliated
unions. Productivity bargaining cannot compensate for other
deficiencies in the factors making for economic growth. It can,
however, make a modest contribution towards greater productivity.

When properly conducted, productivity bargaining has been shown
by experience to provide a number of advantages to workers. In the
first place, it provides a negotiating framework within which unions
can stake a claim for workers to share in the benefits of industrial
and commercial progress. When agreeing to enter into productivity
bargaining the employer is required, in effect, to signify his acceptance
of the principle that workers should benefit from economic expansion
and greater efficiency. In other words, productivity bargaining
expresses an expansionist attitude towards the economy and towards
workers' conditions. A productivity claim thus differs both in form
and content from a claim based on the cost of living or even on
comparisons with the wages or salaries of other workers. A productivity claim reflects and anticipates change. It expresses the view
that workers should have more; more, that is, in the real sense.
Workers are entitled to something better than stability in living
standards.

Another main advantage of productivity bargaining to the unions
is that it brings under scrutiny at the negotiating table a range of
factors contributing to efficiency. The use made of labour is only one
of these factors. Productivity depends, above all, on the kind of

equipment employed, the method of work, and the mechanical power available.

Productivity bargaining underlines the need in many industries to merge the machinery of negotiation and the machinery of consultation. Discussions about the use of resources, productivity, labour practices, and the rewards that should go to labour can then be conducted through a single channel of communication. The base for this machinery should be provided by strong trade union organization and representation at the place of work.

The scrutiny within productivity bargaining of various factors contributing to efficiency leads to a widening of the scope of collective bargaining. Indeed, it is difficult to see how productivity bargaining can be properly conducted unless a wide range of issues, previously not subject to discussion at the negotiating table, are brought into the negotiations. Thus productivity bargaining helps to break down the barrier of certain traditional 'managerial prerogatives'.

A good example of this widening of the scope of negotiations provided by productivity bargaining is in relation to overtime. In the engineering industry there is a long tradition that the decision as to whether overtime shall or shall not be worked is a managerial function. True, there is an overtime and shift-work agreement which lays down a monthly limit on overtime, but the exceptions to this limit for which overtime can be worked are so wide that the agreement is of very limited value. On three occasions in the history of the engineering industry there have been major disputes about overtime. In each of these disputes the unions have sought to establish their right to a say in the extent to which overtime should be worked. Union members resisted the exclusive managerial control of overtime, particularly when in the view of the unions a preferable alternative would have been to provide employment for unemployed members. The last of these national disputes on the control of overtime was in 1922. The unions were defeated. They have, however, always looked forward to the time when the managerial prerogative over the control of overtime could be brought to an end.

In the post-war period the managerial prerogative on the control of overtime has in many firms been eroded. In order to attract or retain labour, firms have been prepared to offer overtime to their work force or to new applicants for jobs. The control of overtime has, therefore, become more a subject for informal arrangements between lower management and workers. Many workers have been prepared to accept substantial overtime because it has given them an opportunity to increase their earnings.

In a number of productivity agreements overtime has either been reduced or even eliminated. This has been achieved by the action of management and union representatives in considering and determining how the various factors or production can be used more efficiently to enable the normal flow of work to be accomplished within the standard working hours. Often new methods and new working arrangements have been introduced to increase productivity. At the same time earnings per hour have been increased to ensure that the take-home pay of workers is not reduced. Through productivity bargaining the control of overtime has been brought under joint regulation. It has been done by considering overtime in relation to other factors affecting the utilization of manpower.

**Benefit to management**

Productivity bargaining has also been of benefit to management. It has enabled management to secure the co-operation of workers in achieving higher productivity. In securing such co-operation management have not found it necessary to advance spurious reasons for seeking the co-operation of workers. On the contrary they have been able to say with justification that greater efficiency would result in benefit to all. Secondly, productivity bargaining encourages management to make a self-critical examination of its own contribution to efficiency. Instead of looking exclusively at the contribution made by labour, management finds it necessary in productivity bargaining to look at the methods and equipment employed. In other words, it is required to look at the control of resources. This is a skilled management task. It is one reason why a good productivity agreement cannot be concluded hastily. Management must look carefully at all the factors making for efficiency.

The third advantage to management of productivity bargaining is that it provides a means for the smooth adaptation of labour practices to changing technology. New production techniques inevitably lead to changes in work content. Sometimes workers may feel reluctant to change their working practices because of fears for the future. They may suspect that change will lead to an undermining of their skill, status, earnings, and security. Productivity bargaining provides a means of taking these fears into account, of discussing them, and of providing safeguards for the future. If certain hardships are inevitably associated with change, steps can be taken to minimize them.

Certain practices concerning the use of labour are based on work methods and technology now out-dated. This applies, for example, in some sectors to the division between skilled and unskilled labour or to the division between process and maintenance work. These divisions have sometimes been defended by unions because workers learnt by experience during earlier years that any departure from traditional practices usually meant the employment of lower-paid labour. Some of these practices were defended by the unilateral action of workers; their unions did not accept that changes in practice—or 'dilution' as it might be regarded—were a matter for negotiation. Within productivity bargaining the adaptation of labour practices to modern technology can be secured to the advantage both of employers and workers.

**Some criticisms examined**

There have been a number of criticisms of productivity bargaining. One of the main criticisms was that the P.I.B.'s guidelines (Report no. 36), by insisting that pay increases under productivity agreements should be granted only in return for the 'direct' contribution of workers, encouraged the conception that productivity bargaining was primarily concerned with the 'buying-out of restrictive practices'. Further, this conception encouraged workers to be restrictive. Every restriction could then have a price tag attached to it. Thus productivity agreements favoured the inefficient and the restrictive because it was they who had the greater scope for raising productivity.

The P.I.B. report answered this criticism by stating that the concept underlying productivity bargaining was the achievement of higher efficiency based on co-operation. It was this co-operation which ought to be rewarded. There were still many cases where it was both possible and desirable to specify required changes in working practice but, said the P.I.B., 'the further development of some productivity agreements has now passed beyond this point and manual workers are increasingly acquiring staff status and staff ~attitudes'.

It was the recognition of the need to relate productivity bargaining to the concept of promoting higher efficiency based on co-operation that prompted an important change in the first of the official guidelines. In Report no. 36 the first guidelines read: 'It should be shown that the workers are making a direct contribution towards increasing productivity by accepting more exacting work or a major change in working practices'.

In Report no. 123 this was amended to read: 'It should be shown that workers are contributing towards the achievement of constantly rising levels of efficiency. Where appropriate, major changes in working practice or working methods should be specified in the agreement'.

This represented more than a mere change in terminology. The new guideline did not speak of workers making a 'direct' contribution. Instead it spoke of workers contributing towards the achievement of constantly rising levels of efficiency. An explanatory paragraph attached to the new guideline rightly put the emphasis on co-operation between management and workers in achieving and maintaining the highest standards in the use of both equipment and manpower.

Nevertheless, the chapter entitled 'Some criticisms examined' in Report no. 123 of the P.I.B. did not break entirely away from the view that what really mattered was the 'direct' contribution of workers. One paragraph, for example, re-emphasized that payment should be related 'to the improved productivity made possible by the contribution of workers and not on the basis of generalized statistics relating to increases in output regardless of whether capital equipment or labour is the source of that increased output'. In another paragraph the report, in referring to the criticisms of the word 'direct' in relation to the workers' contribution to increased productivity, said:

The word was surely aimed at a differentiation between the contribution of labour and the contribution of capital. It is important that this differentiation be maintained and, difficult though it may sometimes be, clearly drawn; otherwise considerably greater earnings could accrue to workers in capital-intensive industries than to those in labour-intensive industries and inequalities in earnings may be enlarged.

Despite the firmness of this assertion that the differentiation between the contribution of labour and the contribution of capital to increased productivity should be clearly drawn the P.I.B. were obviously aware that the problem was not an easy one to resolve. Their report said that they would welcome a further reference from the Government which would help them 'to explore in greater depth the problem of measuring labour's contribution'.

This continuing insistence of the P.I.B. in the text of Report no. 123

on the need to differentiate between the contribution of labour and the contribution of capital—despite the significant change in the first guideline—did not accord with the submitted evidence of the trade union movement. According to the unions the reward to be granted to labour in a productivity agreement should not be confined to the savings made directly by labour. Labour, argued the T.U.C., has a right to share in the *total* savings secured as a result of improvements in production technique, embracing equipment, manpower, and methods.

This view, expressed by the T.U.C., is fundamental to the trade union approach to productivity bargaining. It pointed to an important weakness in the P.I.B. reports on productivity bargaining. Workers are entitled to share in the benefits of increased productivity no matter what the source of the increase may be. An interpretation of productivity bargaining which implied that workers were entitled to share only in the *direct* contribution made by labour would not be acceptable to the unions.

In any case, how would the *direct* contribution of labour be defined? What really matters is the productivity of labour after allowance has been made for the costs of the other factors of production. Labour's contribution consists in its continuous adaptation to new equipment and new methods or, to put it another way, in its adaptation to changing technology.

Another criticism of productivity bargaining which has been advanced among some sections of the left of the trade union movement is that if a union embraces the philosophy of productivity bargaining it is excluded from making claims on other grounds such as, for example the cost of living, the increased profits of an employer or industry, comparability with other workers, or low pay. The plain answer to this criticism is that although there have been many who have advocated that productivity should be a major consideration in many wage claims there are very few, if any, who have advocated that all other considerations should be excluded. No union has found it necessary to conduct all its negotiations on the assumption that productivity is the one and only consideration which is relevant. Each wage claim has to be considered in the light of the circumstances in which it is submitted. The existence of low pay, increased profits, or a labour shortage may be of importance in any particular claim.

Another criticism levelled at productivity bargaining is that it provides wage increases for some workers at the expense of others.

More particularly, so it is argued, wage increases are obtained as a result of the curtailment of the labour force. There are wage increases for some and redundancy for others. If this criticism were to be pressed to its logical conclusion unions would have to oppose every technical change in industry that resulted in fewer workers being required. It is a policy which the trade union movement could not possibly sustain in the long-term. It is based on a static and unrealistic view of the economy. It takes no account of the experience of the growth industries. Technical change can lead to new products, create new demands, and expand employment. The threat of unemployment can be increased more by inefficiency than by technical change.

There are many ways in which the hardships that follow from change may be minimized. On many occasions, even with agreements which provide for a smaller working force to be employed on a new process, it is possible to avoid redundancy. The reduction in the required labour force may be secured by halting recruitment, by natural wastage (e.g. retirement or workers leaving for other jobs), or offering specially attractive arrangements for early retirement or compensation for those who leave voluntarily. In some instances the retraining of workers for other jobs provides a means of smooth transition.

The problem of maintaining full employment cannot in any case be solved by collective bargaining alone. The maintenance of full employment is a task of public policy. It requires a combination of economic measures which fall clearly within the responsibility of the government of the day. The trade union movement can and does seek to influence the government, whatever its political colour, to adopt policies likely to lead to full employment and expansion.

Yet another criticism advanced against productivity bargaining is that some productivity agreements have led to a worsening of workers' conditions. A number of agreements, for example, have provided for the total elimination of tea breaks. That there are agreements of this kind has to be acknowledged. They are the result of bad bargaining. The total elimination of tea breaks is usually counterproductive. Productivity bargaining may produce poor agreements just as any other form of collective bargaining may produce poor agreements.

This same criticism is also voiced sometimes in relation to the wages paid to workers who have concluded productivity agreements. Workers, it is said, may gain an initial improvement as a result of a productivity agreement under which they agree to make concessions but eventually, because of the stringency of the conditions attached to the agreement, the workers' earnings fall behind the earnings of workers employed elsewhere. Again where this has occurred—and there are very few substantiated recorded examples—it is a result of a bad bargain. Under the great majority of productivity agreements wages have risen and continue to remain higher than they otherwise would have been.

The central criticism of productivity bargaining which has come from sections of the trade union and labour movement is that its main purpose is to restore management control over issues on which the authority of employers has been eroded by the growing strength of workers' trade union organization. In other words, productivity bargaining is said by these critics to be a device for limiting the strength and influence of shop stewards. Workers are persuaded voluntarily to surrender their power in exchange for a wage increase. This wage increase may offer a temporary advantage but will soon be overtaken by inflation. Thus productivity bargaining enables management in modern conditions to establish the kind of control which in the years between the wars they maintained by the exercise of managerial prerogatives. Those who carry this criticism to its extreme form argue that productivity bargaining is a means of curtailing the autonomous influence of workers over working arrangements, practices, and conditions.

It is, of course, true that in a productivity agreement a bargain is struck about influence and control. But it is very far from being a one-sided bargain. Management seeks certain changes but in return has to offer improvements in employment conditions and, in addition, has to be prepared to bring within the negotiations matters which previously remained outside. Control is achieved within a context of negotiated consent. This context provides advantages to workers both in conditions and in the scope of collective bargaining. Control within a context of negotiated consent is a very different matter from the unilateral assertion of managerial prerogatives.

If management is to do its job in industry it must have means of control within a framework acceptable to workpeople. It would make nonsense of the role of management in industry if it were to be denied any means of control at all. The vital point for trade unionism is that wherever the control mechanism affects the interests and welfare of workers it must be subject to joint regulation. Productivity bargaining represents an important step in this direction.

**Some problems**

Many as are the advantages of productivity bargaining it would be unrealistic not to recognize that this new development in collective bargaining has brought its own problems. Not all of these problems have been overcome. Some of them in fact cannot be solved within the framework of a productivity agreement. This is one reason why productivity bargaining must always be supplemented by other forms of collective bargaining.

Different industries have different rates of technological progress. There are some service industries where the development of productivity is extremely difficult. Clearly it would be quite unfair if the workers in the industries with a high rate of technological advance were to obtain for themselves the lion's share of the benefit while workers in some service occupations were to receive very little improvement in their conditions. Ideally, rapid improvements in technology should always be accompanied by price reductions except where there are other cost increases, for example in raw materials, which cannot be offset. Price reductions enable the benefit of industrial progress to be shared among a wider public.

There is, however, no means of compelling industries with a high rate of technological advance to share the major part of the benefit with everyone else. Firms will reduce prices if they have reason to believe that by so doing they will expand their share of the market and gain for themselves larger profits by their increased turnover. If they do not expect to make larger profits then they are unlikely to reduce prices.

One of the guidelines on productivity bargaining set by the National Board for Prices and Incomes required that some part of the benefit should be passed to consumers, either in the form of lower prices or as a contribution to stable prices. Clearly this is important. It underlines the relationship between productivity bargaining and a prices policy. It is essential from the point of view of the public interest that the consumer should benefit from changes which lead to higher productivity. If this does not take place workers in industries or services where there are few opportunities to increase productivity will inevitably feel that productivity bargaining is a device to depress their relative standards.

Another problem of productivity bargaining which is sometimes raised by trade unionists is that productivity agreements appear to rest on the assumption that every kind of work can be measured. Trade unionists do not deny that measurement has a part in productivity bargaining relating to some kinds of work, but there are other kinds of work where quality, discretion, judgment, and responsibility form an essential part of the tasks performed by workers. It is much more difficult and in some cases impossible to apply measurement to these qualities.

In its last report on productivity agreements (Report no. 123) the Prices and Incomes Board described work measurement as a technique concerned with the establishment of proper levels of manning so that productivity is brought up to or held at the highest practicable level. The report emphasized that relevant techniques of measurement should be applied to particular tasks. It would be foolish to apply to a drawing office the techniques of measurement which may be suitable for repetition assembly work in an engineering factory. This does not mean, nevertheless, that those concerned with work in drawing offices need give no attention to the relationship between the work to be performed and the manpower employed. The technique of measurement even if it rests on very broad indicators must be relevant to the kind of work undertaken. In this sense—but only in this sense— work measurement is capable of very wide application.

In all areas of employment where work measurement is possible consultation with unions about the techniques to be used and about the validity of the results is essential. It cannot be emphasized too strongly that work measurement does not exclude bargaining; it can be a useful guide to manning standards and can help to eliminate inequalities in work load between one worker and another, but it does not give a scientifically precise result.

Work measurement includes three separate elements; the timing of a range of activities, the assessment of effort by the worker who is being studied, and the estimation of an allowance for fatigue and contingencies. The assessment of effort and the setting of a fatigue and contingency allowance both depend upon subjective estimates. Effort rating and the setting of fatigue and contingency allowances should always be open, if necessary, to challenge by union representatives. Hence it is also necessary that union representatives should receive some training in the techniques of work measurement. To the extent that they are able, if necessary, to challenge knowledgeably the assessments of work study practitioners they will be in a position better to defend the interest of their members.

Just as there is an essential role for collective bargaining and workplace trade union representation in any scheme of work measurement, so too there is need for trade union participation in any scheme of job evaluation introduced as part of a productivity agreement. Job evaluation is the name given to any scheme which provides for the grading of jobs on the basis of a systematic and disciplined study of work content. Most schemes depend on a system of factor rating. The factors may include such requirements as responsibility, experience, skill or arduousness of conditions. A weight is attributed to each factor and each job is assessed by a rating given to each factor requirement. Clearly, the entire exercise depends upon subjective judgments about the qualities required for each job and the weight to be given to each particular quality. Consultation with trade union representatives is essential at every stage in the introduction and operation of such a scheme. Job evaluation schemes tend to be eroded by the introduction of new jobs made necessary by changing industrial circumstances and developing technology. They should be 'audited' at regular intervals with full trade union participation.

When properly applied and periodically 'audited' job evaluation can help considerably to provide for a fairer system of pay differentials related to the contribution required for different jobs. It also introduces a much more sophisticated system of job grading than is provided by the traditional divisions between, say, skilled and unskilled labour or between administrative and clerical employment. When job evaluation is linked to a developed industrial training scheme it can provide new promotion opportunities to men and women who are prepared, irrespective of age or earlier disadvantages, to acquire new skills and to undertake new responsibilities.

**Other grounds for wage claims**

The case for productivity bargaining does not imply that movements in the cost of living and comparisons with the pay of other workers should have no part in trade union claims. Neither can be disregarded by the trade union movement. Movements in the cost of living provide the very minimum for wage settlements. Anything less represents a fall in living standards and no trade union exists to accept that its members' standards should decline. The refusal of unions to accept a decline in workers' living standards is not an expression of obstinacy or ignorance of economic facts but an assertion of the need for, and possibility of, expansion.

The argument for comparability occupies an even more important place in the armoury of trade unionism. The struggle for a common minimum rate for workers performing the same task is fundamental to trade union practice. A common minimum rate, based upon a comparison between equal tasks, eliminates competition between one worker and another. It replaces competition with solidarity. The need for solidarity extends not only to workers with similar tasks in the same enterprise but to workers with similar tasks or similar skill in different enterprises. To argue against this kind of solidarity or comparability is to argue against trade unionism.

In some occupations or industries, for example engineering, the unions have accepted that wage differences based on trade union strength, productivity, the profitability of the enterprise, or local factors can be built on top of a common national minimum rate. In other occupations or industries the unions have sought and have secured a standard rate. This is the case in electrical contracting and for many jobs in the public sector. The establishment of standard rates is also the aim of the day wage agreements in coal mining. Despite these important differences, however, the struggle for a common rate, either as a minimum or as a standard, for workers performing similar tasks or with similar skill, forms part of the history of every union. Comparability, in this sense, is indispensable to trade unionism.

Comparability, however, has also a wider meaning for trade unionists. Trade unionism is not concerned exclusively with occupational interests. It is also concerned with the welfare of *all* who sell their ability to work as the main means of their livelihood. This is not just an expression of altruism, it also expresses self-interest. At different times in the history of the industrial system it has fallen to different sections of the working class and to different unions either to take the main attack in an employers' offensive or to act as the vanguard in a movement for general advance. Thus in the 1920s the miners took the main brunt of the attack on workers' standards. In the 1960s the printing workers led the way in winning a basic 40-hour week. The trade union movement has learnt much from these experiences. The support given by one section of the movement to another, not only in major disputes but also in day-to-day minor strikes or lockouts, illustrates how widespread is the recognition of common interests by trade unionists. The concept of comparability and the recognition of common interest on the part of trade unionists are inextricably intertwined.

There are some services or occupations where it is impossible to base wage or salary claims on measured productivity. There are a number of unions in white collar employment in this situation. In teaching, for example, it would be ludicrous to suggest that the National Union of Teachers should base its periodic claims upon some artificial index of educational productivity per person employed. Inevitably the claims of the teachers will be based on other main considerations such as, for example, the need to offer salaries sufficient to attract and retain an adequate number of suitably qualified persons, and the right of teachers to share in the general improvement in the community's living standards.

Similarly there are other unions representing workers on national salary scales for whom productivity agreements might disrupt an existing satisfactory employment relationship. In the Civil Service, for example, the existence of national scales, coupled with the systematic efforts of the Civil Service Department to increase the efficiency with which labour is employed, provides advantages which outweigh other considerations. The introduction of local productivity bargaining would, almost certainly, create inequities leading to friction and ill-feeling. The salary system would be disrupted. In the Civil Service and in some other public services considerations of comparability will continue to play an important part in salary negotiations.

Arguments about comparability are unhelpful when either they are based on an assumption that a wage or salary relationship which obtained at one period should be maintained for all time, or when they exclude the possibility of discussions about the adaptation of working practices to new techniques. In brief, when comparability is taken to imply that nothing should change, it deserves *always* to be challenged.

**Productivity bargaining not a panacea**

Productivity bargaining is not a panacea for all the problems of industrial relations. Nor can it in all circumstances take the place of other forms of collective bargaining. But where it can be sensibly applied it can help to increase productivity, encourage the adaptation of labour practices to modern technology, increase the earnings of workers, and extend the range of issues subject to the joint regulation of collective bargaining.

During 1970 there was a move away from productivity bargaining. More and more claims were tabled and more and more settlements reached which had little relationship to movements in productivity. Some of these claims and some of these settlements were no doubt prompted by other considerations. The improvement in the balance of payments, the price increases following devaluation, the ending of any real attempt to restrain prices in the private sector, the increases in certain social charges, the need as the unions saw it to stimulate the economy, the low rates of pay in some industries, and the international move forward in wages all helped to give impetus to union demands. Cost–push inflation became a feature of the British economy.

This development, however, is not an argument against productivity bargaining. If the wage increases secured by workers are to result in real improvements in living standards productivity will have to rise steeply. Productivity will then come back as a vitally important factor in negotiations.

# 4

# Industrial training

Of all the resources possessed by society manpower is the most valuable. By his effort and skill a human being is capable of contributing far more to the output of goods and services than the cost of maintaining himself and of acquiring skill. It is from this surplus that capital is accumulated, irrespective of whether the accumulation is made in the name of a landowner, a capitalist, or society at large. Investment in education and training shows a high rate of return.

The working population in Britain is approximately 25 million. For a number of years up to 1966 it had been rising steadily. There were an unusually large number of school-leavers as a result of the post-war increase in the birth-rate, and there was an influx of immigrants. Between 1966 and 1968 this rise ceased, and the size of the working population became stable. In the 1970s it is expected slowly to decline. The reduction in the number of young workers will, however, be much steeper. Moreover, it is not expected that the increase in the number of women workers will continue at the same rate. Annual holidays for manual workers are also expected to lengthen. At present, annual holidays for manual workers are shorter in Britain than in other West European countries. Britain will thus have to depend on increased output per person employed for her economic growth.

Within the total size of the labour force there have been vast changes in the pattern of employment. Before the First World War agriculture employed about a million, mining and quarrying about 700 000, and textiles nearly 2 million workers. Today these industries have shrunk to much less than a half of their former size and the decline continues. In contrast, employment in engineering, construction, distribution, and professional and public services has greatly increased.

The nature of much of the work performed by the labour force has also changed. More and more white collar employees are engaged in industry, commerce, and the public services. The proportion of the total working population employed in manufacturing industry is only slightly more than it was sixty years ago, but even within manufacturing industry the make-up of the labour force has changed. Sixty years ago the manual workers could be divided between craftsmen and labourers and there were few white-collar staff. Today there are large numbers of semi-skilled workers, relatively fewer craftsmen and labourers, and a rapidly growing number of technicians.

The pace of industrial change demands what is described in Sweden as an 'active labour market policy.' Manpower must be educated and trained to meet the needs of a changing industrial structure, unemployment should be kept at a minimum, labour should be employed to maximum social advantage, and provision should be made to minimize the hardship which can fall upon workers when they are required to change their jobs. The danger in the next few years is more likely to come from a failure to respond to this challenge, than from a ready acceptance of the necessity of change. The possibility of economic recession may arise because of a failure to be competitive in world markets. Certainly, inertia would almost certainly hold back the standard of living.

**The need for training**

The trade union movement has consistently supported the extension and improvement of vocational training. Its point of view has been founded, first, on the need for a more skilled labour force as a vital factor in economic growth. The development of skill makes possible higher productivity and provides the basis for rising living standards. Secondly, the unions have been concerned with the advantages of training to the individual worker himself. Training serves his interests. It not only provides him with better earning opportunities, but it can open new interests to him. Work can become something more than physical drudgery.

Training alone, however, is not sufficient. Trained workers need jobs. A policy of full employment, in the sense that there should be a balance between vacancies and workers seeking employment, is essential. The skill of workers needs also to be matched to the available jobs. It would be pointless to train workers for skills which were not required. At any given time some industries are expanding and some contracting. This implies that industrial training must be related to manpower planning.

Moreover the requirements of a satisfactory policy of manpower planning will not be met unless training is regarded as a continuous process for which workers of all ages are eligible. The speed of industrial change today is such that skills acquired during adolescence are unlikely to be adequate for the worker during the whole of his working life. Skills must be developed with changing technology and some workers will find it necessary, with the decline of certain occupations, to be trained for new skills. The unions have supported the principle of adult training and have supported in particular the public scheme operated through Government Training Centres.

Some form of Government-sponsored training scheme has been in existence for many years. The scheme has been used for two main purposes. First, to train people for occupations where skilled labour is short. Second, to provide training for people with special re-settlement needs such as persons made redundant by technological change, disabled persons, unemployed persons, and former members of the armed forces re-entering civilian life. In earlier years the emphasis was on resettlement, but in the 1960s, with the accelerated rate of technological innovation, there was a marked expansion in training facilities for scarce skills. The number of Government Training Centres and training places was more than trebled and a further expansion is now taking place.

The training provided by the Government Training Centres is good. It is based on intensive methods, and a person of average aptitudes is likely to acquire skill more quickly than in many normal apprenticeship schemes. The courses normally last from six to nine months, and thus it is not expected that the skill acquired will be the same as during a good apprenticeship. Nevertheless the groundwork of training is sound and the overwhelming majority of persons trained in Government Training Centres are able to perform valuable work in industry. Their skill develops with further practical experience. Most persons trained in Government Training Centres are accepted by unions. Indeed many unions co-operate in the procedure for the selection of persons for training.

There have been examples of trade union resistance to the employment of adult trainees whose skill was gained at a training centre rather than through a traditional form of apprenticeship but, in general, the trade union movement has been sympathetic to the need to find employment for adult trainees. At the beginning of 1969 a number of unions with skilled workers in the engineering and allied industries gave approval to a Government sponsored scheme for upgrading training in Government Training Centres, providing that unions were fully consulted at local level when an employer wished to take advantage of the facilities. The Government proposed making available up to 400 places annually at Government Training Centres for free sponsored upgrading training. The unions undertook to participate in local discussions about the need for particular skill—bearing in mind long-term requirements and the availability of any unemployed skilled workers—and to examine individually each proposed sponsored worker's case. The unions stressed also that it was important that in organizing the new scheme the Department of Employment and Productivity should keep closely in touch with the industrial training boards established under the Industrial Training Act.

The largest of the unions participating in these discussions on upgrading training, the Amalgamated Union of Engineering and Foundry Workers, accepted the new scheme conditionally. They stipulated that workers who received upgrading training to the skilled level would only be accepted subject to registration under the existing relaxation agreements. They also emphasized that an agreement about a scheme for sponsored upgrading training in Government Training Centres was not to be taken to imply that an agreement covering all forms of adult training had been reached.

This conditional attitude of the A.E.F. towards upgrading training reflects the fear of unemployment of skilled workers, mainly in areas of traditional heavy engineering and shipbuilding. Hence they are disposed to question any proposals for increasing the supply of skilled labour. This is a further illustration of the supreme importance of full employment for trade union co-operation. In the areas where a high level of employment has been maintained for many years and where industry has expanded, principally in and around London and the Midlands, there are many skilled engineering workers who acquired their skill as adults through on-the-job training.

### Trade union support

Because of its support for the expansion and improvement of industrial training the trade union movement, through the T.U.C., has been closely associated with every official public initiative taken since the last war to develop training in Britain. The T.U.C. participated in the inquiry into training by the Joint Consultative Committee to the Ministry of Labour in 1944–5, endorsed the recommendations of the J.C.C. report and supported subsequent efforts to implement them. In 1956 the T.U.C. took part in the work of the Carr Committee to consider arrangements for industrial training in the light of the substantial increase in school leavers then expected and, again, co-operated in activities to implement the findings of the report. The T.U.C. played an active part in the setting-up in 1958 of the Industrial Training Council, through which the unions, private employers, and nationalized industries co-operated to expand training facilities in British industry. The unions were among the strongest supporters of public action for better training and welcomed the passing of the Industrial Training Act in 1964. Unions have played an important part in the training boards established under the Act.

The trade union movement has a distinctive and comprehensive view on training. As already indicated it emphasizes that training must form part of manpower planning, and that full employment provides an essential framework for flourishing training schemes. Equally important is the need to recognize the extremely wide scope for training. Apprenticeship is only *one* form of training. Most adults have not served an apprenticeship and even among youths the majority are not in an apprenticeship or even a formal learnership scheme.

The majority of young people leave school at the minimum age of 15 years or shortly afterwards. Without in any way belittling the importance of facilities for higher education it is necessary to emphasize that public discussion on education and training for young persons tends to be focused disproportionately on the education and training of the minority who study for O and A levels of the General Certificate of Education and who apply for entrance to universities, teacher training colleges, and other higher educational institutions. This is probably inevitable when those who figure most prominently in public discussion are themselves usually the products of the higher educational system and are the parents or grandparents of youngsters who are candidates for or participants in the same system. They are, nevertheless, a minority.

Even among boys who leave school under 18 years of age less than half become apprentices. About one-seventh receive some form of training below the level of apprenticeship. About one-twelfth enter clerical employment, where formal training is unusual, less than

2 per cent go into occupations leading to recognized professional qualifications, and about a third go into occupations without any formal training at all. Even these figures, which leave enormous room for improvement, are better than they were a few years ago.

For girls who leave school under 18 years of age the position is much worse. Less than one in twelve becomes an apprentice—the apprenticeships are mainly in hairdressing—and nearly 40 per cent go into clerical employment where, as with boys, formal training is unusual. Less than 2 per cent go into occupations leading to recognized professional qualifications, and just under 15 per cent enter jobs with some form of training at less than apprenticeship level. About one-third of all girls enter jobs without any kind of formal training.

Among the majority of boys and girls who leave school under 18 years of age there is a vast reservoir of labour which needs to be trained. Clearly there is no likelihood of all these people becoming craftsmen or acquiring equivalent skill. But the idea that workers should be either skilled, semi-skilled, or unskilled, as though everyone can be put in any one of three clearly defined categories, is outmoded. Industry and commerce need an enormous variety of skill and talent at much more than three defined levels. One of the biggest tasks facing the training boards set up under the Industrial Training Act is to encourage employers to provide training for operators among manual workers. Similarly, training is needed for clerical work.

In 1967 in evidence to the House of Commons Estimates Committee the T.U.C. drew particular attention to the need to expand the training of young workers not in apprenticeship schemes. Very substantial advantages would result, said the T.U.C., from 'the introduction and development, with some sense of urgency, of appropriate systematic training for much larger numbers of other young workers from some of whom vocational skills will be demanded at least as exacting as those now demanded of some of those trained by means of apprenticeship'.

In 1966 the Commercial and Clerical Training Committee of the Central Training Committee published a report, *Training for commerce and the office*, which underlined the serious inadequacy of the existing training facilities. Of all office staff under 21 years of age only 8 per cent were being trained and only 7 per cent were receiving day release for further education. Nearly a fifth of all school leavers enter clerical employment, with a much higher proportion among girls, but few employers give serious attention to the training and further education of their clerical staff. In a comment on the report the T.U.C. emphasized the value of planned induction courses for new entrants, the need for adequate supervision of training by employers, and the importance of further education, with paid day release, as an integral element in effective training for office work.

Since the publication of the report the training authorities have sought to improve training methods for clerical work and to extend facilities for it. Some progress has been made but the Third Report of the Central Training Council stated: '. . . many employers are not yet convinced of the value of training office staff, particularly girls'.

## Apprenticeship and modern training

The existence of an apprenticeship scheme provides no assurance that proper training is being undertaken, though great improvements have been made as a result of the work of a number of the industrial training boards. The traditional method of acquiring skill was for the apprentice to learn his trade by working side by side with a craftsman. Systematic training was rarely given. By 'serving time' of five or seven years the apprentice was expected to learn enough to become a skilled worker at the age of 21.

There are many things wrong with such a system. It does not provide for systematic teaching and learning with a programmed course of workshop activity and study. It does not provide for training—or at least part of it—in premises specially equipped for the purpose. It leaves the young worker to acquire skill haphazardly. It does not provide for the integration of training and further education. It does not provide for the incorporation of systematic training in safety as an integral part of industrial training and education. It does not provide training in a form which lends itself to continuous adaptation and flexibility to meet technological change.

To set out the weaknesses of the traditional apprenticeship system is also to indicate the broad lines of needed reforms. These reforms are now being pushed forward by some of the industrial training boards. The record of the Engineering Industry Training Board is outstanding. The basis of the engineering scheme was described briefly in the second report of the Central Training Council.

The Engineering Board has decided that, for the greater part of the first year, craft and technician trainees in all the main metal trades should follow a common broad-based syllabus and that only in the last three months of this first year will any specialization take place. Furthermore, the board has announced its intention that in subsequent years craft skills should be identified by a process of analysis on the basis of a 'module, of the time needed to learn a given skill. The modules chosen for a given trainee will depend on his capacities and the requirements of the employer, and they will not necessarily be within the same 'trade' as previously demarcated. Nor will each module require the same amount of length of training. Thus the total period of training of the craftsman or technician of the future will not be the same in all trades regardless of the skills required. There should be no reason why a recognized craftsman should not undergo further modules to develop or modify his initial training later on in his career. It will no doubt be some years before this new system is generally operative, its implications fully understood, and the problems it raises resolved. Nevertheless, we applaud a bold and imaginative initiative on the part of the board.

The annual reports of the Engineering Industry Training Board have described the progress that had been achieved. The 1969–70 report claimed an overall improvement of 25 per cent in the quantity and quality of training carried out in the industry. The engineering industry now employs some 11 500 full-time training staff. Thousands of apprentices are receiving first year off-the-job training. The module system has been launched, many small and medium firms have grouped themselves together to provide better training facilities, the payment of grants for training operators under the age of 18 has been made conditional on the granting of paid day release for further education, module schemes have been prepared for the training of operators, and many operator training projects have been supervised by the Board. The Board has also encouraged the development of training for office workers, technicians, technologists, supervisors, and management. Altogether over 180 000 employees in the industry, most of whom were craft or technician trainees, were given day or block release.

The further development of the module system of training may ultimately lead to the extinction of the traditional form of apprenticeship system. Already in engineering and shipbuilding the length of apprenticeship has been reduced by one year. The possibility that workers may train as adults in a new but related area of skill underlines the need for continuous learning and adaption to keep pace with technological change.

The work of the Construction Industry Training Board also points to the immense potential advantages of new forms of training in the building industry. In building there are a number of special problems. It is an industry in which the labour force has traditionally been divided between skilled workers and labourers. Yet this rigid division does not correspond with the range of skills required for the most modern building techniques. Prefabricated parts have to be assembled. Only a very small proportion of the woodworkers, bricklayers, masons, and plasterers in the industry are called upon to exercise the highest standards of craftsmanship in their everyday work. This does not mean, however, that they should not be adequately trained for the range of skill which is necessary for their work.

The second problem in the building industry is the need for a more versatile labour force, for workers who possess skill in related trades. This would have a number of advantages. It would make it easier to plan construction work and to eliminate delays caused by trade demarcation. Secondly, it would make it possible for firms to provide more regular employment for their workers. Thirdly, it would provide flexibility to meet changing methods of building construction and, finally, it would be of considerable advantage for maintenance work.

The Construction Industry Training Board has worked out a scheme for training in flexible skills. It provides broad based training for cognate trades, operational training for a particular skill, and modular training for advanced skill. Much will depend on how the industry reacts to these proposals.

The third main problem for training in the building industry is the need to integrate training for skill with training for safety. The construction industry is one of the most hazardous for workers. Accident figures on building sites have increased year by year. There is room for argument as to whether these figures represent a real deterioration in the situation or better methods of reporting, but there is no room for argument on the imperative necessity to make the building industry safer for its workers. The drive for greater safety involves more than training—including, for example, the establishment of joint safety committees at site level, the election of workers' safety delegates, publicity about hazards, safety regulations, the use of protective clothing, and the incorporation of safety factors in design—but training can do much to help. The industry has recently produced a comprehensive accident prevention manual.

The fourth main problem in training for construction arises from the structure of the industry and the widespread practice of self-employment. The industry consists of more than 80 000 firms, of which the great majority employ only a handful of people. This makes effective training difficult to organize. In recent years ten of thousands of workers have become self-employed and building firms have developed labour-only contracting.

The need to encourage adult training is one of the main tasks facing training boards. Training should cover skilled workers whose skill needs to be broadened or deepened to meet the needs of technological change; it should cover workers who need to be trained for new occupations because of the decline of the industries in which they have been employed; it should cover workers who were denied the opportunity to train for skilled work during their youth but who have the ability and aptitude to benefit from training; it should cover operators who require skill but at a lower level than that of craftsmen; it should cover office employment and supervisors; and finally it should cover management. The training should be regarded not as an activity confined to youth and the unemployed but as an activity which continually seeks to improve the skill of the total labour force, and to make it capable of exploiting new techniques of production and service. The idea that the main criterion for eligibility for adult training is unemployment or impending redundancy should go.

**Day release**

For many years the trade union movement has urged that all young persons under 18 years of age should be granted paid day-time release from employment for part-time education. Indeed part-time day release for all young persons was provided for in the Education Acts of 1918 and 1944. Some progress was made by voluntary means but in 1964 a committee, under the chairmanship of Mr. Henniker Heaton, appointed by the Ministry of Education, showed in its report that in 1963 about 550 000 boys and 694 000 girls aged 15 to 18 were receiving no day-time education. The committee said that it was of immediate national importance that the number of young workers granted day release should be increased. Proposals were made for a campaign to extend day release to many more young workers.

The trade union welcomed the Henniker-Heaton report and strongly supported the recommendation that industrial training boards, when drawing up recommendations for training, should pay attention to the need to encourage day release. The Government accepted the main recommendations of the report, and since 1964 education authorities and a number of training boards have co-operated to encourage day release. Nevertheless, the results have not been satisfactory. Some employers and their representative organizations have acted in the spirit of the Henniker-Heaton report. Others have not. In the light of this experience the T.U.C. reached the conclusion that some degree of compulsion on employers would be necessary to provide day release, at least for certain categories of young workers, in addition to the financial incentives provided under the Industrial Training Act.

The 1968 Congress of the T.U.C. carried a resolution expressing concern at the very slow growth in the granting of day release and calling upon the Government to legislate at an early date for compulsory day release for all young workers. Subsequently the General Council decided, in view of the inconclusive results of a lengthy correspondence on the subject with the Secretary of State for Education, to pursue the matter through the Central Training Council. In March 1969 the Central Training Council decided that the Department of Employment and Productivity should write to all industrial training boards asking them to give urgent considerations to a number of proposals to increase the number of young workers granted day release. It drew particular attention to the need to extend day release in clerical and commercial occupations.

**Industrial Training Act**

The Industrial Training Act, which was passed by Parliament in 1964, has three main objectives. It seeks to ensure an adequate supply of properly trained men and women at all levels in industry and commerce; to secure a general improvement in industrial training; and to share the cost of training more evenly between firms. The Act gives the Government the power to set up training boards, which include employers, trade unionists, and other persons, often educationalists. The boards are required, subject to ministerial approval, to make a levy on employers, and they are empowered to make grants to employers who provide training of an approved standard. By the autumn of 1970 boards had been established in 28 industries or related groups of industries, covering over 15 million workers ememployed in about 1 150 000 establishments.

The trade union movement supported the passing of the Industrial Training Act, though a survey of existing schemes, conducted by the T.U.C. some time before, had revealed that nearly all unions replying to the survey were satisfied in general with the existing voluntary national arrangements with which they were associated. The General Council of the T.U.C. took a more critical line. They felt that the existing voluntary arrangements had weaknesses and deficiencies. The survey, they said, 'strengthened the case for introducing some means by which compulsory powers could be brought to bear upon employers to ensure the provision of adequate training'.

Whilst welcoming the Government's new measure for industrial training the T.U.C. criticized what they felt were its inadequacies. Their main criticism was that no provision was made for a strong central authority to direct the work of the training boards. They have continued to reaffirm this criticism, most recently in a memorandum submitted in the summer of 1969 to a committee, appointed by the Secretary of State for Employment and Productivity, to review the functions of the existing Central Training Council. The T.U.C. said that the Central Training Council had neither executive powers nor independence. Some of its functions were not being carried out satisfactorily, notably the considerations of the performance of particular boards, its influence was negligible and its present organization was unsuitable even for its role as an advisory body. In the light of these criticisms the T.U.C. proposed that an 'effective central body is needed to directly influence, oversee and co-ordinate the work of the Industrial Training Boards'. So long as the development of the training systems depended on financial inducements administered by the boards there was a strong likelihood, in the absence of an effective central authority, that each board would tend towards autonomy and be responsible to sectional interests.

The report of the committee, under the chairmanship of Mr. Frank Cousins, was published in April 1970. It recognized that the basic issue before it was the question whether the central national organization for training, above the separate training boards, should be advisory or executive in function. The majority of those who submitted evidence favoured the continuation of an advisory body, though playing a more active role. There was, on the other hand, a strongly held minority view that what was needed was a central authority to replace the Central Training Council and that this authority should have power over the training boards.

The Cousins committee set out the arguments for an executive central authority and said that the case for it was a strong one. The committee felt, however, that there were even stronger arguments against the proposal. Any major reduction in the autonomy of the training boards would, it said, have damaging consequences. The system was still new and constant efforts had to be made to prevent the emergence of hostility and indifference. Training boards needed a substantial measure of autonomy to encourage men of calibre to continue to serve on the boards and for the boards to hold the confidence of their industries.

The conclusion of the Cousins committee was that the advisory character of the Central Training Council should be retained but that certain changes should be introduced—which were described in the report—to enable it to exercise more initiative and influence.

One problem which may emerge in the future, with the continuing autonomy of the separate training boards, is that a new form of industrial rigidity could develop in which occupational skills are related to the needs of one particular industry. There is, however, a growing range of skills, including, for example, those of maintenance workers, draughtsmen, work study practitioners, and computer staff, needed in different industries. A stronger Central Training Council will have to ensure that training for their skills is co-ordinated.

Other criticisms of the training system made from time to time by the trade union movement are that the boards have not always given sufficient attention to the training of operators as well as apprentices; that some boards have not been able to bring about any significant development in induction courses and off-the-job training; that some sections of public service and commerce have not been covered by the new arrangements; and that a great deal more needs to be done in training for office employment. The trade union movement has also continued to press strongly for more day release. Nevertheless, despite these criticisms the unions have been strong supporters of the new initiatives for improved industrial training.

The Labour Government, before its defeat in the 1970 General Election, also made it clear that despite the achievements of the Industrial Training Act there were a number of improvements which would have to be made in the coming years. They were broadly sympathetic to extending the scope of the Act, to reshaping the levy and grants schemes to suit the particular needs of special industrial groups, to meeting the problems of small firms, to the extension of adult training, commercial and clerical training, and management training, and to developing the role of the Government Training Centres.

There has been mounting criticism of the training boards from some sections of employers. It was revealed towards the end of 1970 that the Construction Industry Training Board was in serious financial trouble, and a number of other boards were criticized by employers for being bureaucratic and unresponsive to the needs of small firms. The levy and grant scheme is said by some to be wasteful. A major review of the system is being undertaken by the Department of Employment. Changes will be needed in the light of experience, but it would retrograde if the net effect were to curtail the attention now given to industrial training.

The record of the T.U.C. on industrial training is a very good one. Of all the representative organizations in Britain it has been among the most vigorous in its advocacy of better and more extensive industrial training, particularly for the majority of citizens who do not benefit from higher education and who do not enter an apprenticeship scheme. The record of a number of individual unions catering primarily for skilled workers has not been as good or as forward looking as that of the T.U.C., but even among the majority of craft unions the predominant influence has been towards better and more systematic training.

# 5

# Restrictive practices

To its critics trade unionism is synonymous with restrictive practices. Its main purpose, its most severe critics say, is to interfere with the free and proper functioning of the labour market. The unions when at their strongest, so the argument runs, restrict entry into occupations in need of labour. They prevent the employment of workers at rates of pay which many workers would be prepared to accept. They insist on drawing artificial boundaries around particular occupations so that certain work is reserved exclusively for members of a particular union. They impose limitations on output, insist on inflated manning scales, resist the introduction of new methods if the privileges of the existing workers are threatened, and discourage or even prevent the employment of women in occupations for which they are physically suitable. To put it briefly, the unions contribute to industrial inefficiency. They impede the optimum use of productive resources, including both men and equipment.

It is probably only a relatively small minority who take this extreme critical view. They regard trade unionism as a major handicap to efficiency. There is a larger body of opinion, however, which accepts that trade unionism has brought advantages to society, but that these advantages have to be balanced against the disadvantageous restrictions which were developed in conditions of mass unemployment and which persist to this day but with very much less justification. In its evidence to the Royal Commission on Trade Unions and Employers' Associations, for example, the Confederation of British Industry said:

Underlying most restrictive practices is the fear of redundancy and unemployment (although it is doubtful whether under present day conditions this fear is justified or whether, in fact, such practices can do other in the long run than create the unemployment they are designed to avoid).

## The regulatory function

The answer to these charges is not to be found in a denial of the regulatory functions of trade unions within the labour market. This market, without collective bargaining on the part of workers, is one in which in most circumstances there is great disparity in strength between the buyer and the seller. The employer is in a very much stronger position. By combining together in a union the workers can help to redress the balance of economic power.

There will, of course, always be scope for argument as to whether a particular form of regulation provides an apparent immediate benefit at the expense of longer-term improvement. However, even where it can be shown that labour has introduced a restriction which cannot be justified it does not invalidate the general proposition that, in the light of industrial history, labour must to a very substantial extent look to its own strength to regulate employment conditions to ensure that it shares in the benefits that flow from economic expansion, and that it is protected against many different kinds of hazards.

The regulatory function is, therefore, at the very core of trade union-ism, and the regulation of the labour market is impossible without some restrictions. Every collective agreement on minimum wages restricts the freedom of employers and workers to enter into a contract of employment for a lower rate of pay. The same, in substance, is true of collective agreements on the length of the basic working week, on minimum holiday entitlements, and on overtime premium rates of pay. Restrictions may also be imposed to protect the safety, health, and welfare of workers. Thus in foundries there are restrictions to curtail dust as a protection for workers against silicosis. In all manufacturing industries there are restrictions against exposed moving machine parts which could endanger the fingers and limbs of workers. More recently, as a result of the Offices, Shops, and Railway Premises Act, restrictions have been introduced on office conditions for the protection of office workers. All these are examples of restrictions that are justifiable and, indeed, are necessary. These socially necessary restrictions on pay and hours of work, and for safety and welfare may be imposed either by the State, by collective agreements between employers and unions, or, in the absence of legislation or collective agreements, by the unilateral decision and enforcement of groups of workers.

Nor can it be said that there is some kind of difference in principle between a restriction imposed by the State, a restriction imposed by collective agreement, or a restriction imposed unilaterally by a group of workers. A restriction imposed unilaterally by workers, say on hours of labour, may become at a later stage the subject of a collective agreement and at a later stage still may be enforced by legislation. The purpose of the restriction does not become socially defensible only from the moment that it is embodied in legislation.

**Socially defensible?**

Once it is accepted that restrictions on the use of labour are necessary the real point at issue is to distinguish between practices that are socially defensible and those that are not. A research paper prepared for the Royal Commission on Trade Unions and Employers' Associations defined the term 'restrictive labour practice' as an 'arrangement under which labour is not used efficiently and which is not justifiable on social grounds'. This definition points to the importance of social judgements in assessing particular restrictive practices. This puts the emphasis in the right place. Generalizations about restrictive practices, other than the important generalization that some restrictions on the use of labour are essential, are not helpful.

The Confederation of British Industry in their evidence to the Royal Commission on Trade Unions and Employers' Associations said that a restrictive practice was any work practice, collectively operated, which hindered or acted as a disincentive to the more effective use of labour, technical skill, machinery, or other resources. They went on to state that the problem posed by restrictive practices was essentially 'one of over-manning, or under-utilization of manpower resources'. They instanced over-manning resulting from union reluctance to accept manpower economies when new methods or machines are introduced; demarcation rules between workers of different crafts, skill, or grade which were increasingly unrealistic in relation to present day production methods; systematic time-wasting to hold down bonus earnings, to create the need for overtime, or as a sanction to support a claim for higher pay; resistance to technical innovation; resistance to work study; resistance to the recruitment of apprentices; reluctance or refusal to undertake overtime or shift work; and resistance to the idea that merit rather than seniority or union membership should be the main criterion for promotion.

The C.B.I. claimed that these restrictive practices affected most industries in one form or another, though they were associated in the public mind mainly with the docks, shipbuilding, ship-repairing, and newspaper printing. The Confederation said that restrictive practices had seriously retarded economic growth in the past, and that 'they represent perhaps the greatest obstacle to future growth'. The C.B.I. also claimed that restrictive practices were conducive to restrictive or obstructive attitudes among unions and workers. This was specially serious when unions were resistant to technical innovation.

Despite the strength of its criticisms the C.B.I. did not condemn all restrictive labour practices. Some, it said, reflected a desire—genuine even where misconceived—to maintain craft standards or to promote safety and health. In others, such as the regular resort to overtime, the employer may have been a consenting, if not a willing, partner. Underlying most restrictive practices, said the C.B.I. was the fear of redundancy and unemployment.

The Engineering Employers' Federation also gave evidence to the Royal Commission on practices which they alleged prevented the most effective and efficient use of labour. Among the practices they described were the insistence by some district committees of unions on the use of skilled labour on machines which could quite easily be operated by semi-skilled labour; the limitation of production and earnings to a predetermined level; the widespread refusal to recognize anyone as skilled unless he had served an apprenticeship beginning in his juvenile years; the limitation on the number of apprentices for training for a particular trade, irrespective of the need for skilled men; and disputes between unions over which craftsmen should carry out certain tasks.

**Trade demarcation**

One of the restrictive practices most frequently complained of is that of trade demarcation; namely, the insistence by a union that a certain range of work is to be carried out exclusively by its members, irrespective of whether other workers could equally well do the same work; and, conversely, the insistence by a union that its members should not be called upon to undertake certain tasks which fall outside their normal range of duties, even though they may be perfectly capable of undertaking these other tasks.

Demarcation is associated with the division of labour; and the division of labour was developed and is still being developed in many circumstances as a means to higher productivity. It does not, therefore, always follow that labour specialization is maintained only at the insistence of unions. Indeed, there are a number of examples from industry where the unions rather than the employers have sought to establish one grade for craftsmen instead of a number of grades related to different levels of skill. In the industrial civil service, for example, the unions pressed for the abolition of a grading scheme with two grades of craftsmen even though they agreed to such a scheme when it was first introduced. In shipbuilding where 'who does what?' disputes are more frequent than elsewhere it was the employers, as the late Lord (Ted) Hill used to insist, who first broke down the comprehensive skills of the boilermaking craft and required men to specialize as platers, riveters, caulkers, drillers, markers-out, and welders.

The real problem arises when developments in technology and in methods of production outgrow the existing division of labour, and when simultaneously it seems likely that there will be a fall in the demand for the labour of a particular group of workers or for a particular trade. The problem is accentuated when this takes place in an industry where there have been frequent periods of heavy unemployment or prolonged industrial strife. Hence the particular problem in shipbuilding and ship-repairing.

Shipbuilding is not, nevertheless, typical of other industries. In the great majority of firms changes in production methods take place without serious demarcation problems. Smooth change is more typical of British industry that the 'who does what?' disputes which capture the newspaper headlines. Even in shipbuilding there have been far-reaching changes in building techniques since before the war. Welding has replaced riveting, flame-cutting is now extensively used, and heavy sections are prefabricated in workshops. The labour force has adapted itself to the new methods.

If there is to be flexibility coupled with the advantages of the division of labour the main source of improvement for the future will be found in better methods of training. The new training methods in engineering and construction, under their respective training boards, will help enormously for the future. The emphasis needs to be on the training of workers with a wider range of basic skills so that they are adaptable to industrial needs and changing methods.

One very interesting and helpful development of recent years is in electrical contracting. In the summer of 1966 an agreement was reached between the National Federated Electrical Association and the Electrical Trades Union on proposals for increased productivity, pay increases over a period of three years, the re-grading of electricians according to skill, the elimination of craftsmen's mates, and the setting up of a Joint Industry Board. Under the scheme, which has since been continued and developed, the Joint Industry Board is concerned with all aspects of industrial relations and productivity. It took over apprentice registration and carried through the re-grading of workers. Now nearly all new entrants, including most current apprentices, will have to train for relevant qualifications, secured not only by experience but by passing examinations of the City and Guilds of London Institute. In reporting on the scheme the National Board for Prices and Incomes stated: 'Although we have reservations on individual features of the Agreement, taking it as a whole we think that it represents a notable achievement in industrial relations, training, and employment policies over a remarkably short period'.

By eliminating the grade of 'mate' the scheme gave the industry the opportunity to get away from the traditional concept of 'pair' working, that is to say allocating all work to a craftsman and his assistant. The case studies made by the Prices and Incomes Board revealed that the practice of 'pair' working was very deeply entrenched. The report added (Report no. 120, paragraph 65): 'It is probably true to say that the men have been quicker to adapt than managements . . .'

Many productivity agreements have also provided for changes in the traditional demarcation between occupations. Report no. 23 of the Prices and Incomes Board, which examined productivity agreements in Esso, electricity supply, I.C.I., British Oxygen, and Alcan Industries, stated (paragraph 12): 'Flexibility or interchangeability between different groups of workers is a feature of all the agreements mentioned in the reference made to us. In other words certain groups of men agree to perform tasks previously the exclusive concern of another group.'

A further report of the P.I.B. (Report no. 36) stated that flexibility had not always been easy to achieve despite the formal terms of productivity agreements. Nevertheless progress had been made. This finding was confirmed in a later report, Report no. 123, on productivity agreements.

One industry where trade demarcation does add unnecessarily to costs is printing. Again the evidence is supplied by reports of the P.I.B. Report no. 2 of the Board said that the Board had been given many examples of practices which had the effect of limiting productivity. They included the insistence by separate unions that certain tasks—even very menial tasks—were to be carried out only by their members.

The P.I.B. said, however, that the belief that these practices all spring from sheer perversity on the part of the workers was unfounded. The employer 'must carry his share of the responsibility for failing to remove the sense of insecurity which accounts in large measures for the workers' attitude'. The unions, the P.I.B. report pointed out, had demonstrated their willingness to negotiate progressive agreements with progressive employers. In return for higher productivity, including realistic manning, such agreements provided higher pay, longer holidays, full pay during illness, and a staged retirement plan.

Some four and a half years later in a report on national newspaper costs and revenue the P.I.B. said that there was no doubt that the industry was making an effort to achieve more realistic manning levels. It described the progress which had been made. The report also spoke of the need to extend mixed manning agreements in machine rooms on the lines of an agreement between the Society of Graphical and Allied Trades (SOGAT) and the National Graphical Association (N.G.A.) concluded when the *Sun* was transferred from the International Publishing Corporation (with N.G.A. machine managers) to the *News of the World* (with SOGAT machine managers). The P.I.B. report said that the economic problems of the national newspaper industry did not arise solely, or even principally, from labour costs or trade union resistance to technical innovation, but productivity bargains which substantially reduced labour costs per unit of output might, nevertheless, make the difference between survival and extinction for an individual newspaper.

### Restrictions on entry

In most industries unions do not impose restrictions on the entry of new workers, though they may reserve the right to make representations to employers about the ratio of apprentices to skilled workers. This is not just an expression of their concern about a possible surplus of trained labour in the future. It is also an expression of concern about the use of cheap labour and the debasement of training standards. Though there *are* examples of unions holding the number of apprentices to a level lower than could be justified there are many other examples of youths, both boys and girls, being employed as a source of cheap labour with very little regard for their future skill and employment as adults.

There are some declining industries, notably the docks, where very strict control is maintained on the recruitment of labour. This control is certainly supported by, and in some cases may have been prompted by, the unions. It is a form of control which is socially necessary if industries are to contract with minimum hardship to those already employed in them.

Dr. W. E. J. McCarthy (1964) estimated that just under three-quarters of a million workers were employed in one form or another in pre-entry closed shops. By this was meant employment in which the worker has to belong to or be accepted by a particular union before he can be engaged by the employer. Among newsprint workers in London, for example, the unions control the supply of labour: workers are engaged through the unions. On the docks and for seamen in the Merchant Navy employers and unions form a pool through which workers are recruited. The pool is confined to workers accepted by the unions. In general printing there are many firms which recruit skilled labour only from among union members trained as apprentices. The practice sometimes also obtains for craftsmen in shipbuilding, in iron and steel, in a number of engineering firms, and among maintenance craftsmen in a variety of industries. In iron and steel, according to Dr. McCarthy, promotion among process workers depends on seniority, with the proviso that the promotion opportunity will be vetoed by the union if the candidate for promotion is not, at the time the vacancy arises, a fully paid-up union member.

Dr. McCarthy discussed at considerable length in his book whether or not the closed shop in the variety of forms it takes is justified. He reached the general conclusion that it *was* justifiable, although it is sometimes used unnecessarily and is liable to abuse. It was usually demanded, he explained, where unions faced problems of organization and control which without the closed shop would remain insoluble.

Under the Conservative Government's Industrial Relations Bill any agreement for a closed shop will be void except in very special circumstances. The Government's view, set out in its Consultative Document on the Bill, is that 'an employer should be free to employ anyone who has the necessary skills'.

There are strong arguments against such a sweeping provision. Control of entry in some sections of employment is essential for stability. A pre-entry closed shop for trained persons, with regulated entry and a post-entry closed shop for persons accepted for training, provides this control. The arrangement often suits both employers and workers. A pre-entry closed shop may also help to provide more stable employment and stable industrial relations in sections where there is intermittent employment. It is unwise to generalize either for or against a closed shop.

### The exclusion of women

A practice or tradition for which there is much less justification is the exclusion of women from certain kinds of work for which, *given training*, they would be suitable. This practice rests primarily on the traditional subordinate status of women in society and the consequent restriction of opportunities for them to acquire skill.

The denial of opportunities to girls and to women is by far the biggest single restrictive practice in the use of labour in our society and it is one for which the trade union movement is not primarily responsible even though union members, under the influence of traditional social attitudes, have sometimes sought to perpetuate it. Nevertheless, in its attitude to the right of women to greater opportunities at work the trade union movement has a better record than some of its critics.

A survey of women in industry was prepared by Miss Nancy Seear for the Royal Commission on Trade Unions and Employers' Association. It was subsequently published as one of the Royal Commission's research papers. It pointed out that only 29 per cent of women manual workers in industry were ranked as skilled (figures for 1966). If the clothing and textile industries are excluded the women's position worsens markedly. In the engineering and electrical group of industries, for example, only 13 000 women were ranked as skilled although the industry employed a total of no less than 357 640 women.

Miss Seear pointed out that in non-manual employment the most outstanding feature was the overwhelming concentration of women in clerical and other office work and their virtual exclusion from managerial, scientific, and technical work. Out of 897 120 women employed in non-manual occupations just over 2 per cent were working as scientists, technologists, technicians, or draughtsmen. Of those employed in clerical work only a very small proportion were receiving any form of planned training. Much of the work was semi-skilled or unskilled.

There are three main arguments sometimes advanced to explain or even to justify the poor job opportunities available to women and hence the waste of their talent. The first is that because of marriage and childbearing women do not stay long enough in employment to justify expensive training. Clearly there is some weight in this assertion but not as much as is often assumed. Many women do not give up their employment on marriage, though they usually do so on the birth of their first child. More and more women, however, return to employment later in life, either in their late thirties or early forties. Thereafter they have a working life which may span anything up to 20 years. Thus the working lives of many women are interrupted but not terminated by childbearing.

The second main argument which is used to explain the poor job opportunities for women is that they are much less reliable in attendance than men. This is the kind of generalization which, as Miss Seear pointed out in her research paper, does not stand up to detailed investigation. Some women are unreliable but so too are some men. Moreover, as many employers are aware, many married women who return to work in their thirties, forties, and fifties are exceptionally reliable and conscientious.

It is in any case necessary to take a broader social view of the contribution of women in employment. Although there have been changes in domestic habits in the direction of a greater sharing of household duties between men and women it is still the case that much the greater part of the responsibility of running a house and caring for the family falls on women. There is as yet inadequate provision of facilities, such as child-care centres or nursery schools, to lighten the domestic burden. The need to protect women from excessive labour in employment arises as much from this social tradition as from their weaker muscular power. Thus the case for greater opportunities for women, and, incidentally, for equal pay for equal work is not contradicted by the need for protective legislation for women in employment. During the time spent in employment the abilities of women should not be wasted.

The third argument sometimes used against the employment of

women in skilled work is that for many kinds of jobs their abilities and attributes are of a different order from those of men. Again, this is a generalization which over a very wide range of employment cannot be sustained. It is, of course, true that women generally have less muscular power than men and that, conversely, they are often more dexterous. But for many jobs muscular power is not the prime requirement. For these jobs there is no natural or inherent reason why women should be less suitable than men.

In her paper Miss Seear argued that since training was the key to opportunity, education and training should be the first line of attack. It was urgent that girls schools should be helped to attract qualified staff—if necessary on a part-time basis—to teach mathematics and science. In addition, better vocational guidance and training facilities were required for older married women who returned to employment. There was need too to increase the opportunities for the part-time employment of women. Attention was also drawn to the low pay of many women workers.

### Resistance to new methods

An allegation which is sometimes made against unions is that all too frequently they oppose new methods and the introduction of new machinery. Their purpose, it is said, is to preserve the jobs of existing union members and to combat anything which might threaten job security. Union members are accused of being Luddites for acting with the same motivation as the machine breakers among the hand-loom weavers and other textile workers in the early days of the Industrial Revolution.

The answer to this charge is that over much the greater part of industry and commerce it simply is not true. Even in the firms or organizations where examples of resistance to new methods can be quoted there is nearly always another and deeper explanation. The failure in industry, where it has occurred, to introduce modern methods and machinery is attributable in the great majority of cases to weaknesses in structure—for example, the existence of too many small firms with a shortage of capital—or to a lack of professional skill in management rather than to opposition from workers.

In the technologically advanced industries, where if workers' resistance to new methods and machinery existed the effect would be most apparent, there is very little evidence of any obstruction. In the motor-car industry, electronics, chemicals, steel, and oil refining, co-operation has been secured. Even in the shipbuilding industry it would not have been possible to revolutionize the methods of ship construction over the last 35 years, with the development of welding, flame-cutting and prefabrication, without substantial co-operation from the workers.

This denial that resistance to the introduction of new machinery is characteristic of trade unionism does not imply that changes in methods of work *always* proceed smoothly. Any proposed change in methods of working is likely to have a direct impact on workers' interests. It may affect the pace of work, intensify labour, increase boredom, affect earnings, call for new skills or more flexible working practices, require different working hours, or threaten job security. New methods and new machinery are also likely to increase productivity and thus they have an important bearing on negotiations about wages and conditions.

Disputes can arise on any or all of these issues. If as a result of one of these disputes workers in a particular establishment refuse to accept the introduction of new methods until a settlement has been reached concerning the employment arrangements it does not necessarily indicate they are resistant to change. Their concern is to ensure that when change takes place they share in the benefits, that the change is made in an orderly manner, and that any inevitable hardships are minimized. It would be utterly unrealistic to expect workers to press for anything less. Indeed, the weakness in British industry has not been that workers have had too much say in determining the circumstances and consequences of change; it is rather that they have had too little.

When viewed in this way the examples which are sometimes quoted to illustrate the contention that unions are obstructive assume a different significance. The introduction of shift work to replace day work may, for example, alter fundamentally the way of life of a family. It cannot and ought not to be assumed that just because it is more economical to work very expensive capital equipment on two shifts or even continuously rather than for, say, 8 hours each day from 8 a.m. to 5 p.m. with one hour for lunch, that workers who have been employed on day work ought immediately to respond to the need for shift work. The change-over must be negotiated patiently, and the new working arrangements should bring benefits to workers to offset the accompanying inconveniences. The benefits may include, for example, higher pay, shift allowances, a shorter working week, and longer holidays.

## Job security

Concern about job security is fundamental to the trade union response to industrial change. A great deal depends upon the economic environment. If there is full employment and confidence that it will be maintained workers are more likely to be receptive to innovation. If in such an environment an industry is expanding and is able to provide steadily improving wages and conditions to its workers it is not usually difficult to secure workers' co-operation to the introduction of new methods and machinery.

In declining industries, particularly with an ageing labour force, the problem is more difficult. The experience of the coal-mining industry shows, however, that even in a period of contraction and even with a labour force which has few other employment opportunities in the immediate neighbourhood, it is possible to press ahead with an ambitious programme of mechanization. Nearly all Britain's coal output is today cut and loaded by mechanical power. Only a few years ago the greater part of the output was 'hand got'. This change in the method of coal mining has been accompanied by similarly far-reaching changes in the wages system.

The experience of the coal mining industry surely shows what it is possible to achieve, even under the most difficult circumstances, given an imaginative and sympathetic attitude on the part of management and a readiness on the part of government to recognize that there is a social obligation to meet some of the cost of assistance to the men.

In addition to pursuing policies for full employment the government of the day can also help by other measures to lessen the fear of insecurity. The Redundancy Payments Act and the provision of centres for training and re-training are examples of this kind. They are designed to reduce some of the hardships of change and the fears which go with it.

## Limitation of output

Another restrictive practice, about which allegations are often made, concerns the limitations of output by workers and deliberate over-manning. In the research paper on restrictive labour practices prepared for the Royal Commission on Trade Unions and Employers'

Associations examples of such practices were given from printing, heavy electrical engineering, shipbuilding, and the docks.

There can be no doubt that there *are* examples in British industry and commerce of deliberate over-manning and restrictions on output. Deliberate over-manning cannot be justified except where it can be shown that it is a temporary arrangement to retain a team of labour during a slack period or in other special circumstances. Both employers and unions may be fully justified in some circumstances in retaining labour on a particular job in excess of their immediate requirements. Long-term manpower planning may, indeed, require that on occasions particular jobs should be over-manned. It would suit no one's purpose, and least of all the national interest, if a team of labour was discharged during a temporary slack period and then another team had to be recruited when business became brisker. Finally an employer might decide to retain certain older workers, to be employed at less than normal potential, as part of a deal to meet an anticipated contraction of the labour force following the introduction of new methods and machinery. It cannot, therefore, be assumed that deliberate over-manning is always wrong.

Some restrictions on output are also open to different interpretations. A target, the achievement of which may demand substantial effort, may also serve as an output limit. This is not uncommon on piece-work in the engineering industry. To describe this arrangement as a restrictive practice is to rob the words of any real meaning. A group of workers, or an individual worker on batch production in an engineering machine shop, may set a target of a given number of pieces per working day. To achieve this target may demand substantial effort. There may, however, be an understanding between the workers that the number of pieces should not be exceeded. The target figure for the number of pieces to be completed is usually related to an observed limit on piece-work earnings. Thus, in the language of an engineering workshop, all workers may be required by mutual understanding not to exceed 'double time'. This limit is imposed because of the very real fear that, despite the sustained effort necessary to reach it, any excess in earnings will result in moves by management to cut piece-work earnings. This would not usually be done by a straightforward cut in the piece-work price—though even this is not unknown—but by changing the method, including the tooling, materials, or machinery. The job would then be retimed.

Thus both deliberate over-manning and restrictions on output ought not to be condemned without regard to the circumstances in which they operate. If, however, they are devices to create more jobs than are reasonably necessary, taking into account the fluctuations in labour requirements and the social need to minimize the hardship of change particularly for old or handicapped workers, they ought not to be defended. Such deliberate waste of labour is not, nevertheless, characteristic of British trade unionism though it certainly exists in some areas of employment. Where over-manning exists—and it is fairly widespread—it is more usually due to inefficient working arrangements, defective organization, and poor planning.

**Working arrangements**

A bigger problem than deliberate over-manning is the existence of working arrangements which, though not thought of primarily as over-manning, represents an inefficient use of labour. The arrangement that a mate should always accompany a craftsman on certain jobs is an example of this kind of inefficient practice. So too is the demarcation between trades in some industries, or the demarcation between all process work and all maintenance work in others. In engineering the insistence in some firms that men who have not served an apprenticeship should never have the opportunity to be upgraded is another wasteful practice. Some of these inefficient working arrangements could be changed fairly quickly; others require for their removal an extension of training facilities. Forward-looking trade union leadership can help to overcome local resistance to change. Co-operation has to be obtained by explanation and by the existence of conditions which inspire confidence.

Both work study and job evaluation can be used to bring about better and more efficient manning arrangements in industry and commerce. When they are used in conjunction with collective bargaining on wages they can also provide information to assist negotiators in formulating a wage structure related to the needs of the job. But work study and job evaluation, as already pointed out in an earlier chapter, do not remove the need for collective bargaining. Management, unions, and work study practitioners should accept from the outset that both work study and job evaluation can be employed satisfactorily only within a framework which provides for consultation, negotiation, and a measure of joint control.

## Overtime

There is substantial evidence that some of the overtime worked in British industry—particularly where regular long hours of overtime are worked—is a reflection of the inefficient use of resources. In its evidence to the Donovan Commission, for example, the C.B.I. said that 'a good deal of overtime is worked for non-productive reasons'. In his summary of Mr. Whybrew's research paper on *Overtime working in Britain*, prepared for the Donovan Commission, the research director, Dr. McCarthy said: 'It is concluded that overtime patterns in Britain can best be explained in ways which have little or nothing to do with production demands'.

The Donovan Commission itself confirmed that one of the frequent inefficient practices in the use of labour was regular substantial overtime.

The Prices and Incomes Board, in its detailed study of overtime in Report no. 161, took a less emphatic view. It showed that actual hours of work for men in manual employment in Britain were higher, except perhaps for France, than in any other Western industrialized country at a comparable stage of development.

British arrangements, it said, had advantages in that it had given a wider variety of choice to employers and workers than would have been possible in a situation where overtime was more tightly regulated, The P.I.B. report accepted, however, that 'where overtime for men is high, it is much easier to find examples of industries and undertakings in which there is much scope for improving efficiency than where there is little apparent scope.' The existence of high levels of overtime, said the P.I.B., is therefore in itself sufficient to constitute a warning that resources are perhaps being wasted and that remedial action is necessary.

Since the war the normal hours of work in Britain, i.e. the length of the basic working week, have been reduced for nearly all manual workers and for a large majority of white collar workers. For manual workers the length of the basic working week has been cut in most industries from 48 or 47 down to 40 hours. This reduction took place in successive steps and it was not until the 1960s that the 40-hour week became general for manual workers. There are still some exceptions, notably agriculture, but the great bulk of manual workers today have a basic 40-hour week.

For white-collar staff the length of the basic working week is frequently shorter. Many have a basic working week of $37\frac{1}{2}$ hours, though there are still some who work side by side with manual workers and have a 40-hour week. On the other hand, there is a growing number of white-collar staff, particularly in the London area, who have a basic working week of 35 hours.

The reduction in the basic working week has not, however, led to a corresponding reduction in the actual hours of work. On the contrary, with the reduction in the basic working week there has been an increase in the average amount of overtime worked, particularly for adult men. This applies more in manual employment than among white-collar staff. Approximately two-thirds of adult men manual workers work overtime. The average number of weekly hours worked by men manual workers, excluding sickness, is about $46\frac{1}{2}$. These are not much less than in 1947–8. They rose again in the 1950s to reach a peak of nearly 49 hours in 1955. Since that time they have dropped again, with some fluctuations, to their present level.

The significant fact about these figures is the extent of systematic overtime in Britain. In his research paper on overtime working Mr. Whybrew said, 'an examination of patterns of overtime working in greater detail suggests that it is the same industries, firms, and individuals who work high overtime'.

The evidence that regular substantial overtime, or systematic overtime as it is sometimes called, is not necessarily related to production needs is strong. Following the introduction of the well-known productivity agreements at Fawley, agreements have been made in a considerable number of firms to eliminate overtime without any reduction in take home pay. The work has been reorganized to ensure that the level of output is maintained. Where agreements of this kind have been carefully planned they have been successful. The reports of the Prices and Incomes Board on productivity bargaining have shown that though not all agreements have been successful there have been sufficient number to establish beyond reasonable doubt that it is possible to eliminate a good deal of overtime in British industry without any decline in output. A number of P.I.B. reports covering specific industries or sections of industry have also pointed to examples of artificial overtime.

The question that arises is why does systematic overtime persist on such a widespread scale despite the evidence that much of it is inefficient? To many managements faced with an output problem overtime appears to offer a simple, relatively cheap, and immediate solution. It is simple because there are usually plenty of workers ready to work overtime. Moreover, the introduction of overtime does not make any demanding calls upon managerial skill. It enables a manager to give an impression of special effort even though he may have done little or nothing to reorganize the resources of production to increase productivity. Overtime appears to provide a relatively cheap way out of output difficulties, because even though premium payments have to be made to labour for work done in overtime hours there are certain overheads which are not increased. It is always arguable that in the short term overtime provides an economical way to expand output. Another advantage which overtime appears to offer to management is that its effect on output appears to be immediate. When overtime is introduced output goes up. It is only after an interval of time that the pace of work becomes adjusted to the longer hours.

All this does not imply that all overtime is unnecessary. Obviously if there is an emergency in an undertaking as, for example, with a breakdown of machinery or power supplies, overtime may be essential for the necessary repairs to be completed. Temporary overtime may also be essential to meet certain seasonal peaks of activity. Sometimes steps can be taken to secure a more even flow in demand from one period to another, but in some industries seasonal fluctuations are unavoidable. Agriculture is a case in point.

There are also certain industries or occupations where a certain amount of scheduled overtime is almost unavoidable if adequate services are to be provided at unusual hours. Perhaps the most obvious example is that of the transport industry, though even in transport undertakings care can be taken to limit, as far as possible, demands for scheduled overtime. Another group of workers who may be required, unavoidably, to work overtime are those engaged on the maintenance of plant. It may be necessary for them to start work somewhat earlier or to finish their work somewhat later than the majority of the labour force. Maintenance work may also have to be done during week-ends or over shut-down periods.

One reason often advanced by management for the introduction of overtime is that is it necessary because of a shortage of labour. In some cases, the reason may be valid, but all too frequently overtime once introduced is then maintained without any simultaneous effort to reorganize the work, to reduce the demand for highly skilled labour, or to train new people in higher skills. The shortage of labour becomes an excuse for regular overtime.

In many establishments overtime is maintained—whatever formal reasons may be given for it—not so much for production reasons but to provide a more acceptable level of earnings to workers. In other words, overtime becomes a substitute for a pay increase. Once the habit of overtime is established it creates its own expectations. Workers become accustomed to higher earnings and understandably resist any attempt to reduce them. If these expectations are not then fulfilled through higher basic rates strong pressure is exerted for regular overtime. Employers are often willing to accommodate this pressure and offer overtime as an inducement to attract and retain a labour force.

The relationship between overtime and low basic rates of pay is not an automatic one, but the relationship undoubtedly exists. In general, low-paid workers work more overtime than highly-paid workers, time workers work more overtime than piece-workers, and unskilled workers work more overtime than skilled workers. There are exceptions to these generalizations and it certainly does not follow that in any individual case a high basic rate will lead to a reduction in overtime, or conversely that a low basic rate will be accompanied by high overtime. Nevertheless, these generalizations taken over the greater part of industry are valid. They point to the existence of a relationship between the level of pay and the amount of overtime.

Employers who offer overtime as a means of increasing earnings, even though they may be aware that a good deal of nominal working time is wasted, are able to satisfy themselves that they are not thereby breaking collective agreements on wages or creating problems for other employers by offering higher than normal basic rates. Employers who offer high basic rates but who eliminate overtime are more likely to be criticized by their fellow employers than an employer who provides the same earnings but who does it by artificial overtime.

Overtime once introduced very quickly becomes habitual. It easily becomes an irreversible practice. The effect on management is usually bad. Regular overtime leads to managerial inertia because it discourages initiative. The desire of many workers, particularly adult men, to work overtime to supplement their earnings stands in contrast to policy declarations of the trade union movement deploring systematic overtime. Unions are well aware that the existence of low rates of pay provides only part of the explanation for high overtime in Britain. There are many higher paid workers who also look for regular overtime. The plain fact is that to a number of workers regular overtime has become a way of life which they are reluctant to change. The desire for leisure occupies a low place in their scale of preferences.

Most of the industrialized Western European countries have means whereby overtime is brought under regular scrutiny and whereby some restraint upon it is imposed either by law, public agency, or collective agreement. In Britain there are a number of collective agreements which impose restrictions on the amount of overtime that may be worked. They nearly all, however, provide for so many exceptions that their effectiveness is much reduced. Moreover, most of these agreements do not impose an absolute limit on the amount of overtime that can be worked. They impose a limit on the amount which may be worked without the local union organization being consulted.

There is also a strong tradition among employers in Britain that the control of overtime is, and should continue to be, regarded as a managerial prerogative. As the Donovan Commission pointed out, 'the effective regulation of overtime has been hindered by the notion of managerial prerogative' (paragraph 93).

In conditions of high employment coupled with the tradition of managerial prerogative it is inevitable that in practice the level and distribution of overtime should be determined informally in the workplace. Because of the denial of any effective role for the unions in the control of overtime there are few, if any, restraints on the readiness of substantial numbers of workers to accept overtime. The decision to offer overtime is also frequently made by first line supervisors and lower management. All too often top management whilst continuing to insist in words on a managerial prerogative has in practice lost control of overtime. Far from being a retreat it would be a step forward for them to establish the joint regulation of overtime. There are a number of influential employers who recognize this reality of industrial life.

To bring about a significant change in the wasteful use of overtime will be no easy matter for Britain. It will demand not only a change in the habits of many workers but a change in the thinking of many employers. The control of overtime carries with it important implications for management. It requires that management should use its skill to ensure, as far as possible, that productive resources are so organized that the required output can be secured within the basic hours of work. It requires also that plans should be made to reduce wherever possible the fluctuations in the flow of work and to arrange for the supply of raw materials and bought-out parts in the right form in the right place and at the required time.

The control of overtime also carries with it important implications for the unions. It requires that unions should be prepared to participate in the machinery of control. It also requires that unions should seek constantly to persuade their members of the disadvantages of systematic overtime.

Probably the most effective way of reducing overtime in Britain would be to encourage the drawing up of more productivity agreements designed to deal with overtime within a context of promoting greater efficiency. Overtime could then be considered not in isolation from, but in relation to, the more effective use of resources, including manpower, equipment, and methods.

**Conclusion**

It would be pointless to deny that there is considerable inefficiency in the use of labour in British industry and commerce. Most of this inefficiency is attributable to traditional working arrangements and shortcomings in the organization of work. Restrictive practices enforced by trade unionists also play a part but, as the Donovan Report rightly pointed out, 'most restrictive labour practices are not enforced by the unions as such'.

The way forward is through negotiations for the better use of all the factors of production. Workers will expect to share in the benefits that flow from this greater efficiency. Their co-operation will be much more forthcoming if the economy is expanding and if there is full employment.

# 6

# Implications for trade union structure and functions

Technological change has important implications for the structure and functions of the trade union movement. These implications are all the more significant when, as in recent years, the pace of change has been accelerating. Some of the changes which are having the greatest impact on trade unionism can be readily distinguished. They include the breakdown of traditional crafts and dividing lines between occupations, the development of new techniques calling for different kinds of skill, the rapid growth of white-collar employment, the increasing importance of large multi-plant firms, and the development of new methods of wage payment and man management which have contributed towards shifting the focal point of collective bargaining in many manufacturing industries towards the workplace.

It is these changes, together with the certainty of continuing change, which makes some of the traditional forms of trade union organization inadequate for modern needs. Craft identities have been undermined by technological change. In many industries there has been a tremendous growth in both the number and proportion of semi-skilled operatives. With the use of new materials different crafts have merged one into the other or have been replaced by new techniques. Some of these new techniques require not only semi-skilled operatives, but also a very highly skilled minority who are part craftsmen and part technicians.

Trade union organization based on the identity of traditional crafts cannot cater for these developments without changing its form. This change may take the shape of an amalgamation between related trades, as with the boilermakers, shipwrights, and blacksmiths; or the woodworkers, bricklayers, and painters; or the sheet metal workers, coppersmiths, and heating engineers; or the electricians and plumbers; or the compositors and printing machine-minders.

Alternatively it may take the shape of a broadening of membership to embrace not only skilled workers from related crafts but semi-skilled and unskilled workers who are employed together in the same establishment or sector of industry. Today, for example, the Amalgamated Union of Engineering Workers recruits workers of all levels of skills employed within the engineering and allied industries or on engineering maintenance work in other industries.

There is, however, yet another significant development in forms of trade union organization. It is the growing readiness of a number of unions to become 'open-ended' in their scope of recruitment. They are prepared to recruit wherever their day-to-day activities take them, with little regard for traditional industrial or craft boundaries. These boundaries are felt to correspond to a technology which is becoming increasingly outmoded. These are many examples to illustrate this new trend. With the increasing importance of synthetic fibres the distinction between textile and chemical manufacture is becoming blurred, The importance of plastics is affecting many different industries including engineering, building construction, and pottery. The use of computers provides a new occupational identity for technical staff in many branches of industry and commerce. All these changes have affected trade union organization. The faster growing unions in Britain are those which have adopted the 'open-ended' approach to recruitment. Both the Transport and General Workers' Union and the Association of Scientific, Technical, and Managerial Staffs tend to follow this general principle.

**Multi-unionism**

The wider adoption of 'open-ended' recruitment policies is likely to increase rather than to diminish the problem of multi-unionism for workers employed in a single establishment. The Donovan Commission report distinguished between two kinds of multi-unionism. The first is the situation in which each of the main occupational groups in an establishment—different crafts, process workers, supervisors, clerks, technicians—are organized by different unions. The second is the situation where there is more than one union competing for membership within a given group of workers. The first of these two multi-union situations is the more frequent in British industry and is often found in industries employing different kinds of craftsmen. The second situation occurs more usually among semi-skilled process workers or among the unskilled. There is, however, a growing problem of multi-unionism among some groups of skilled workers, including engineering technicians and supervisors.

Multi-unionism has a number of disadvantages. It can lead to demarcation disputes and it can be destructive of trade union co-operation. Impediments to technological change are more likely in multi-union situations than in industries or establishments where there is but one union. It is sometimes argued, on the other hand, that competitive trade unionism acts as a spur and that in its absence unions may become complacent and authoritarian. Undoubtedly, there are examples of unions becoming complacent and authoritarian, with a self-perpetuating leadership, in situations where they have an assured membership without fear of challenge from any competing organization. The real solution, however, does not lie in the promotion of inter-union competition. Such competition can soon degenerate into warfare with each union entrenching itself in an established position and becoming resistant to change. The solution to complacency, ineffectiveness, and dictatorial methods in trade unions must come through rank and file activity. Change can be brought about within unions as a result of pressure from below. There are a number of examples of such changes in Britain in the post-war years.

Much has already been done by the trade union movement to deal with the problem of multi-unionism. More remains to be done. It has often been argued in the past, and it is still sometimes argued today, that the only solution is through industrial unionism, that is that there should be one union for all employees in an industry irrespective of occupation. On at least three occasions—the first during the mid 1920s, the second during the early 1940s and the third during the 1960s—the T.U.C. General Council investigated possible changes in trade union structure and reached the conclusion that it would not be practicable to reorganize the British trade union movement on industrial lines. Their consistent view has been that there is no simple form of organization which is suitable for all circumstances, and that the diversity of circumstances within industry, as well as between industries, implies some diversity of trade union structure.

In a report to the 1964 Congress the General Council of the T.U.C. described at length why, in their view, industrial unionism did not provide an answer to the problems of trade union structure. In the first place there was no consensus of opinion as to what constituted the boundaries of an industry. Industries could be defined differently

7*

for different purposes. Secondly, technological change modified conceptions of what constituted an industry. This implied that any scheme of trade union reorganization on industrial lines would need to be revised from time to time if it were not to lead in the future to difficulties similar to those it was designed to correct. Thirdly, industrial unionism did not provide a satisfactory basis of trade union organization for workers in a particular craft or occupation who often felt a stronger tie with similar workers in other industries rather than with workers in their own industry who follow a different occupation. When workers with a particular skill move from one industry to another they can retain membership of the same union if the union is based on craft or occupational identity. Fourthly, there are some industries that are too small to sustain a separate union. Large unions are in a better position to provide specialized services. A more practicable alternative to a small industrial union is to have a general union with membership spread over several smaller trades or industries but with trade or industrial sections which are largely self-governing.

The conclusion of the T.U.C. was that it was neither desirable nor practicable to prepare a central plan for trade union reorganization based upon any one principle of organization or any particular theory of trade unionism. The aim rather should be to assist unions to come together in amalgamations and closer working arrangements with a view to eliminating inter-union competition, providing more effective services, making a greater impact on wider industrial and national problems, and extending trade union membership.

Following this report the General Council of the T.U.C. convened a number of conferences of groups of affiliated unions to discuss the possibilities of amalgamation and closer working arrangements. A new impetus was given towards structural change within the trade union movement.

## Changes in structure

A significant number of changes in trade union structure have taken place in recent years. In shipbuilding the boilermakers, shipwrights, and blacksmiths amalgamated to form one union. In engineering the Amalgamated Engineering Union, foundry workers, draughtsmen, and constructional engineers formed one union but with substantial sectional autonomy. The sheet metal workers, coppersmiths, and heating enginers formed one union from three separate predominantly craft unions. The supervisors and scientific workers came together to form the Association of Scientific, Technical, and Managerial Staffs (A.S.T.M.S.) which then grew very rapidly and attracted to itself unions catering for medical practitioners and insurance staff. The electricians and plumbers amalgamated to create a strong and expanding union with interests in building, electricity supply, engineering, and maintenance work. In building the woodworkers, painters, and bricklayers formed a long overdue amalgamation for the construction industry. In printing there were a number of amalgamations leading to the formation of two major unions, the Society of Graphical and Allied Trades (SOGAT) and the National Graphical Association (N.G.A.). Subsequently, however, there was serious dissension in SOGAT and it seems possible that the constituent unions will revert to their own identity.

Despite these changes there is still need for further far-reaching reform. On the railways, in engineering, building, printing, and textiles, to mention only a few of the more important industries, there are too many unions. The T.U.C. in their recommendations have pointed a way forward. Discussions for further change are proceeding among a number of unions.

The changes that are likely to take place will not, however, bring to British industry a tidy pattern of trade union organization. Nor will they establish industrial unions even in industries where it would be relatively easy to determine the main boundaries. The reason is the simple one that existing unions are not prepared to change fundamentally their basis of organization. This is true not only of the general unions, but also of many craft unions and of unions whose main base is in one industry but who also organize related groups in other industries.

Given therefore that radical reorganization is extremely unlikely in the foreseeable future, but that the process of amalgamation will continue with large unions straddling different industries, one of the main tasks will be to improve working arrangements between unions. Competitive multi-unionism can be reduced by agreements between unions for 'spheres of influence'. This is particularly important at the level of the workplace. Unions ought to be able to agree that in each establishment only one of them should in future recruit workers of a particular grade. The Donovan Commission suggested that the T.U.C. should adopt the principle of 'one union for one grade of work within one factory' and should work towards it not by trying to enforce the principle upon unwilling unions but by initiating co-operation where it thought that its intervention might be acceptable to the parties.

### Disputes between unions

One of the by-products of the controversy between the Labour Government and the T.U.C. concerning proposals for changes in industrial relations outlined in the White Paper, *In Place of Strife*, was that the General Council of the T.U.C. were given additional responsibilities by affiliated unions to assist in resolving disputes between unions. Such disputes were said to fall, wholly or mainly, into four categories.

(1) A difference about which union should be recognized (recognition).

(2) A difference about the union to which a particular group of workers should belong (membership).

(3) A difference about which union members should carry out particular work (demarcation).

(4) A difference about the policy which should be pursued in respect of terms and conditions of employment (wages and conditions).

The solution of inter-union disputes about demarcation and wages and conditions usually depends not only on the unions directly involved but also on subsequent agreement with an employer. The General Council of the T.U.C. are prepared in such cases to make a recommendation for the solution of the dispute and if the unions are prepared to act on it and if the employer indicates that he is prepared to accept the T.U.C.'s findings the General Council will then make a binding award. In inter-union disputes about recognition and membership, the T.U.C. are prepared to make a binding award if discussions do not lead to a solution which is acceptable to the unions concerned.

In their relations with each other, unions are required by the T.U.C. to be bound by the so-called Bridlington Principles and Regulations. In making awards in inter-union disputes concerning membership the T.U.C. Disputes Committee observe the requirements of these principles and regulations. The Bridlington Principles relate to proposals adopted by the 1939 Congress and are intended to minimize the likelihood of disputes between unions and particularly those disputes which result from inter-union competition. These proposals were themselves a clarification of principles of inter-union relationships enunciated by the 1924 Hull Congress.

Following the 1969 Congress the General Council conducted a survey of joint working agreements between affiliated unions. Some 139 agreements were submitted by 40 unions, and another 15 unions stated that they had no written agreements at all although most of them pointed out that there were many well-observed but undocumented undertakings between themselves and other unions with whom they were in frequent contact. Altogether the replies covered unions representing a total membership of nearly 6 700 000.

The General Council also analyzed 221 inter-union disputes reported to them and 92 demarcation disputes reported to the Department of Employment and Productivity over the previous ten years. The analysis showed that the largest number of disputes occurred in engineering, followed by construction, steel, shipbuilding, printing and paper, and food, drink and tobacco. A number of conclusions were drawn by the General Council from the analysis. The first was that disputes concerning the trade union membership of supervisors, foremen, clerical, and technical workers appeared to be growing. The second was that in engineering, steel, construction, and shipbuilding the disputes concerning demarcation of work arose in some cases from technical changes which blurred former craft boundaries. The third was that many unions had recently taken an increasingly wide view of their areas of recruitment and had extended their activities into new industries and processes. Certain of these areas of difficulty were not covered by inter-union agreements.

Arising from this review the General Council of the T.U.C. wrote to all unions in the summer of 1970 reminding them of the advantages of developing closer working arrangements with other unions. They also expressed readiness to assist in promoting co-operation between unions.

Meanwhile the T.U.C. General Council, and particularly Mr. Victor Feather personally, had been increasingly active in dealing with disputes reported to them in accordance with the changes in rules adopted following the controversy over *In Place of Strife*. The 1970 report of the General Council stated that in the twelve month period ending in June 1970 over 180 disputes had been reported to the T.U.C., some by affiliated unions, some by the Department of Employment and Productivity and some by employers seeking direct assistance from the T.U.C. In about two-thirds of the reported disputes there was an actual or threatened stoppage of work. Nearly half of the reported cases concerned inter-union problems. The T.U.C. claimed that in about 30 cases their intervention helped to avoid a stoppage of work, and that altogether the number of days work saved by T.U.C. intervention was, at a rough guess, between 2½ and 3 million.

Critics of the trade union movement have pointed out that despite T.U.C. intervention the number of days lost through industrial disputes in 1970 showed a sharp rise over previous years. This, however, is not a fair criticism of the role of the T.U.C. in carrying out the undertaking which it gave to the Labour Government following the withdrawal of the controversial proposals in *In Place of Strife*. The responsibility for disputes about wages, conditions, recognition, and victimization rests elsewhere than with the T.U.C. Its special responsibility is to assist in reducing and eliminating inter-union disputes. In that respect the T.U.C. General Council was justified in claiming to have made a significant contribution towards averting or curtailing disputes in a number of industries.

In a report to the 1970 Congress entitled *T.U.C. structure and development*, the General Council recorded that unions themselves were placing emphasis on the need to avoid the haphazard growth of multi-unionism. Inter-union rivalry, it was pointed out, should be eliminated in order to avoid the waste of trade union organizing resources and to prevent friction between unions. The changes in T.U.C. rules in 1969 represented, in the opinion of the General Council, an endorsement of the view that 'failure by unions to reconcile their disputes peacefully was a matter of concern to the movement as a whole'. The General Council said that they intended to examine, in the light of their experience, whether the Bridlington Principles or the powers or practices of Disputes Committees needed to be changed.

### Industrial committees

An interesting further development foreshadowed in the General Council's report on *T.U.C. structure and development* was the establishment of a number of industrial committees which would act as a channel through which unions and the General Council could co-operate in identifying and solving major problems affecting particular sectors, and through which T.U.C. initiatives could be taken in relation to collective bargaining developments affecting groups of unions. Industrial committees on these lines, under the auspices of the T.U.C. have already been established for steel and for the construction industry. Industrial committees would be able to facilitate union co-operation on such matters as recruitment, inter-union relations, collective bargaining objectives, safety, and the activities of the relevant Economic Development Committees and the Industrial Training Boards.

The establishment of industrial committees would have implications for existing trade union federations—such as, for example, the federations which now exist in building, printing, engineering, and shipbuilding—and these implications would in due course have to be considered by the unions. If industrial committees, closely associated with the T.U.C., were to become effective they would probably render redundant some of the existing federations. In any case the views expressed to the General Council by affiliated unions showed that many of them considered that the functions and activities of the existing federations and the trade union sides of National Joint Councils needed to be reviewed. The indications are, therefore, that machinery for trade union co-operation in multi-union situations will develop increasingly under the auspices of the General Council of the T.U.C.

This also is part of the significance of the establishment by the General Council of a Collective Bargaining Committee. It developed from a former General Council committee set up to deal specifically with incomes policy. The Collective Bargaining Committee will be able to bring together cognate groupings of unions to discuss common problems of collective bargaining and, if desirable, to consider a collective bargaining strategy. Discussions of this kind are likely to lead to greater trade union co-operation.

### White-collar employment

The growth of white-collar employment as a result of industrial and commercial change is having a very important effect on the composition of the trade union movement. Well before the end of this century the number of white collar workers in Britain will exceed the number of manual workers. Traditionally, trade union organization has been stronger among manual workers than among white-collar staff, and it is in some of the most strongly organized industries, including coal-mining, cotton textiles, shipbuilding, and railways, that the decline in employment has been steepest. It is this decline, together with the expansion of the more weakly organized white-collar occupations, which resulted in a near stagnation of total trade union membership for a large part of the post-war period. If anything, the proportion of the total labour force organized in trade unions showed a slight tendency to decline, though probably this trend has now been halted and reversed with the much more recent general expansion of trade union membership.

Trade union membership amongst white-collar employees is distributed very unevenly. In public employment, including the Civil Service, local government, teaching, and the nationalized industries, trade union density among white-collar employees is high. The unions into which they are organized are effective and amongst the most efficient in the whole British trade union movement.

In private industry and commerce the density of trade union membership among white-collar staff is very much lower than among white-collar staff in public employment. Professor George Bain has estimated the density of white collar unionism in the public sector of the economy as over 80 per cent whilst in the private sector he put it at slightly over 10 per cent (Bain 1970). Even this 10 per cent, however, was spread very unevenly. Among shop assistants, for example, there is one area of employment, the co-operative movement, where trade union membership approaches 100 per cent. Outside the co-operative movement trade union organization among shop assistants is confined to a number of multiples and department stores. In many small shops trade union organization is non-existent.

Among clerical workers in private industry and commerce trade union organization is generally weak, though trade union membership is now growing fairly rapidly in the engineering industry. In banking the National Union of Bank Employees has been able to develop effective organization over the greater part of the field, and after a struggle lasting many years has made important gains in establishing negotiating rights. In insurance some of the outside staff dealing with the collection of premiums from the homes of subscribers are well organized, but among the inside staff effective organization has, until recently, been limited to a minority of companies.

In contrast to the generally low level of trade union membership in the white-collar private sector there are, nevertheless, three or four groups who have traditionally been well organized. They include journalists, draughtsmen in engineering and shipbuilding, musicians, and some of the other employees in entertainment.

The white-collar union which has made the most spectacular progress in membership is the Association of Scientific, Technical, and Managerial Staff (A.S.T.M.S.). This union was formed from an amalgamation of two unions, ASSET, based originally on foremen, and the Association of Scientific Workers. When they amalgamated their membership began to leap forward almost immediately. A.S.T.M.S. have adopted a very vigorous 'open-ended' recruitment policy and it has paid off handsomely. They have extended trade union organization into many fields of white collar employment and have provided the organizing facilities to meet a very obvious need.

Unlike many other countries Britain has succeeded in maintaining one national trade union centre, embracing not only unions based on different organizational principles and unions with declared political objects and unions without such objects, but also unions covering both manual and white collar workers. This feature of British trade unionism is extremely significant and, mistakenly, is often taken for granted by British commentators outside the trade union movement.

For many years there have been no serious tendencies making for the disruption of this unity. On the contrary, the T.U.C. has become more and more representative of the whole of the British trade union movement. The two largest white-collar unions, the National and Local Government Officers' Association and the National Union of Teachers, are now both affiliated to the T.U.C. There are only two white-collar unions of significance which are still outside the T.U.C. but neither sees itself as being in opposition to the mainstream of trade unionism. The two unions are the Institution of Professional Civil Servants, representing civil service professional and technical staff, and the Society of Civil Servants, representing civil service executive grades. One section of the Institution of Professional Civil Servants, formed from the former Society of Technical Civil Servants representing draughtsmen and other drawing office technical staff, is affiliated to the T.U.C., and it is no secret that many of the leaders of the I.P.C.S. are sympathetic to joining the T.U.C. Much the same is true of the Society of Civil Servants.

Many of the white-collar unions, including both the I.P.C.S. and the S.C.S., though not the National Union of Teachers, are also affiliated to the National Federation of Professional Workers. This, however, is in no sense a rival national centre to the T.U.C. It works in harmony with the T.U.C. and is represented on the Non-Manual Workers' Advisory Committee of the T.U.C.

Each year the T.U.C. holds a special conference of unions catering for non-manual workers. The 1970 conference was the thirty-third such occasion and it was attended by 125 delegates from 48 unions. It discussed a wide range of issues affecting white collar staff, including the effect of technological change on office employees. A special report on automation in offices was approved and resolutions were adopted, among other matters, on training, productivity agreements, and the health hazards of shift workers.

The annual conference of T.U.C. non-manual unions elects a number of representatives to a Non-Manual Workers' Advisory Committee. They are joined on the Committee by a number of members of the General Council of the T.U.C. The report on *T.U.C. structure and development* stated that the General Council believed that there were valid reasons for maintaining both the conference of non-manual unions and the Non-Manual Workers' Advisory Committee. The report said that through technological change many of the previous distinctions drawn between manual and non-manual workers had been breaking down. Nevertheless, the occupational interests and attitudes of non-manual workers were sufficiently distinct to require special arrangements. This did not imply, however, that the interests of non-manual workers were so distinctive that they could be hived off completely for separate consideration. This, said the T.U.C., would harm the development of broadly based overall trade union policies.

According to T.U.C. figures at least 28 per cent of all affiliated members are non-manual workers. At the end of 1968, 42 unions with a membership of 1 588 300 said they catered exclusively for non-manual members; a further 24 unions said they had both manual and non-manual members. The non-manual membership of these 'mixed' unions was 603 000. Thus at the end of 1968 the number of affiliated non-manual workers was 2 191 300. Since that time there has been a significant growth in non-manual trade union membership and, in addition, in May 1970 the National Union of Teachers became affiliated to the T.U.C.

The blurring of the distinction in some occupations between manual and white-collar employment and the desire of unions to extend membership into occupations which are growing has led to some difficulties between a few white-collar unions and unions catering traditionally for manual workers. A.S.T.M.S. has been involved in a number of disputes, particularly with unions in engineering and steel.

The more traditional unions claim the right to organize groups of technicians, supervisors, and sometimes clerical staff but they have not usually been able to establish negotiating rights for them, particularly in engineering. The British Iron, Steel, and Kindred Trades Association (BISAKTA) has been more successful in steel. A.S.T.M.S., on the other hand, has negotiating rights in engineering wherever it can establish majority membership in a grade (providing no other union has negotiating rights), but it has no national negotiating rights in steel. The Clerical and Administrative Workers' Union has negotiating rights for its members in engineering.

### Workplace bargaining

The growth in the proportion of workers employed in large establishments, the need to provide negotiating arrangements for dealing with problems which arise in the workplace, the further application of new managerial techniques, and the strengthening of trade union organization have all contributed to the development of workplace bargaining during the last thirty years. This too has implications for the structure and functions of the trade union movement.

In the first place, it requires that there should be a system of trade union workplace representation. In their evidence to the Donovan Commission the T.U.C. said that there were at least 200 000 workplace representatives, known under a variety of titles such as shop steward, father of the chapel, or office representative.

Secondly, the role of the workplace representative needs to be properly integrated within the structure of the union. There are no hard and fast rules for this which can be applied to every union. The important requirement is that the union's structure should recognize the workplace representative and that his functions should be defined. There should be no ambiguity about his role and his relationship with, say, the members, the branch, the district committee, and the local full-time official of the union. Improvements have been made in recent years in a number of unions to clarify and strengthen the role of workplace representatives.

Discussions have also taken place in recent years in a number of unions on the relationship of the branch to workplace representatives. There has been a trend towards identifying the branch with a particular workplace, group of workplaces, or sector of industry. In some unions this has always been the custom. A number of unions have also adopted the practice of calling conferences or meetings of workplace representatives from the various establishments of big firms or combines.

The existence of shop stewards' committees in factories and other workplaces where a number of stewards have been elected is also helpful for promoting effective trade unionism. If the stewards all belong to one union difficulties are unlikely to arise unless the stewards come into conflict with decisions taken by higher committees of the union. On the other hand, shop stewards' committees which embrace stewards from a number of unions can put themselves in difficulty if they fail to observe the requirement that stewards are ultimately responsible to their individual unions and not to a multi-union shop stewards' committee. Co-operation in a multi-union shop stewards' commitee depends on goodwill, tolerance, and a willingness to use available channels for bridging differences between unions and reaching agreement on a common policy for collective bargaining purposes.

In September 1970 the Transport and General Workers' Union suggested in a letter to the T.U.C. that shop stewards' committees should be represented on trades councils. At present trades councils, which exist under the control of the T.U.C., consist of representatives drawn from affiliated branches. The T.G.W.U. proposal, if adopted, would represent a significant departure from tradition, but is in line with the new emphasis on 'grass roots' democracy. It would also recognize that in many areas the real focal point of trade union interest and activity has tended to shift from the branch room to the place of work. Mr. Jack Jones, the General Secretary of the T.G.W.U., was reported as saying that the movement needed to be strengthened 'down the line', and that he took the view that trades councils as local organs of the T.U.C. should have more contact with factories.

The importance of workplace representatives in modern industrial relations requires that they should be properly informed about union rules, policy, and the content of agreements. They should also be helped to discharge their functions effectively. Their functions include recruitment and the strengthening of organization, negotiation on matters which fall within the scope of workplace collective bargaining, ensuring that agreements are observed, the dissemination of information about union activities and, in some unions, the collection of subscriptions. If all these tasks are to be performed competently training is nearly always essential.

There has been tremendous expansion in the training of workplace representatives in recent years. Many unions have issued booklets for the guidance of representatives and many now arrange training classes or schools for their stewards. Some unions have established residential training colleges where stewards can attend courses, with full or partial pay, lasting a week, a fortnight, or even a month.

The T.U.C. has also greatly expanded its training facilities. The T.U.C. Training College provides courses both for voluntary and full-time union representatives and officials. In addition, the T.U.C. organizes a number of residential summer schools, provides postal courses, and arranges regional week-end schools, day schools, and classes on industrial relations subjects. The total number of students participating in these T.U.C. schemes now approaches 25 000 per year. The T.U.C. also gives advice to Industrial Training Boards on the training of workplace representatives, and in 1969 and 1970 it co-operated with the B.B.C. in planning and preparing a series of educational television broadcasts designed to help workplace representatives in productivity bargaining. It was estimated that for the autumn 1969 series at least 7000 workplace representatives took part in study groups formed with the co-operation of management to view the series during working hours.

Despite this great expansion in training there are still tens of thousands of workplace representatives who have received no formal training whatever for their union duties. A survey conducted for the Donovan Commission, for example, found that only 30 per cent of stewards said that they had taken part in training courses. The figure would probably now be higher, but there is still much to be done.

The growth of large-scale industry and the important part played by workplace union representatives has brought to public attention the extent to which the actions of one group of workers may affect the employment of thousands of others. Strike action in one area can disrupt production in other areas and cause temporary unemployment among far more men and women than those involved in the initial dispute. The report on *T.U.C. structure and development* said: 'It is also essential that groups of members should have regard to the consequences of their actions for other trade unionists, and should accept that membership imposes limitations on their right to take unilateral action.'

These words of caution do not imply that disciplinary measures or legal sanctions provide the answer to strikes which are unconstitutional or which directly affect the employment of other workers. It can never be assumed that strikers are necessarily wrong; their grievance may be real and there may have been a failure to deal with it expeditiously or justly. They may also have been subject to provocation. On the other hand, the strike weapon can be used unwisely and not only as a last resort and ultimate sanction. The need is to improve negotiating procedures so that changes affecting workers' interests are not introduced without discussion and so that grievances of all kinds can be raised within the procedure, dealt with quickly, and resolved at workplace level.

**The future**

No one would argue that the existing structure of the British trade union movement is ideal. If it could be planned anew its profile would look very different from what it now is. Even then problems would begin to arise with changes in the structure of industry. None of the countries which are often quoted as having a better trade union structure than Britain have been able to overcome all difficulties. Both in Sweden and Germany, for example, where unions are more usually organized on industrial lines, there are nevertheless exceptions, and in neither country has it been possible to maintain one united national trade union centre embracing the overwhelming majority of organized manual workers and white-collar workers. Critics of British trade unionism are inclined to concentrate on the problems which arise from the structure of the trade union movement but fail to see that in practice it has proved to have certain advantages which other countries, with an apparently tidier and more logical trade union structure, would wish to emulate.

The real question, therefore, is not whether the British trade union movement can be radically reshaped, but whether the process of change and adaptation from within the existing structure is sufficient to take account of changes in industrial technology and the ever widening aspirations of trade unionists. The answer can be cautiously optimistic. The reports of the T.U.C. show clear recognition of the need for continued progress, and the changes in trade union leadership have made the movement more alert to the wishes of the membership than it was some twenty years ago.

# References

**Chapter 1**

BROWN, E. P. and BROWN, M. A. (1969), *A century of pay*. MacMillan, London.

**Chapter 2**

EVELY, R. and LITTLE, I. M. D. (1960). *Concentration in British industry*. Cambridge University Press.

MONOPOLIES COMMISSION (1969). *General observations on mergers*. Annexe to report on proposed merger between Unilever Ltd. and Allied Brewers Ltd.

**Chapter 5**

MCCARTHY, W. E. J. (1964). *The closed shop in Britain*. Blackwell, Oxford.

**Chapter 6**

BAIN, G. S. (1970). *The growth of white collar unionism*. Clarendon Press, Oxford.

# Index